ROUTLEDGE LIBRARY EDITIONS:
FOOD SUPPLY AND POLICY

T0187524

Volume 1

POLITICS AND POVERTY

POLITICS AND POVERTY

A Critique of the Food and Agriculture Organization of the United Nations

JOHN ABBOTT

Routledge
Taylor & Francis Group

LONDON AND NEW YORK

First published in 1992 by Routledge

This edition first published in 2020
by Routledge
2 Park Square, Milton Park, Abingdon, Oxon OX14 4RN

and by Routledge
52 Vanderbilt Avenue, New York, NY 10017

Routledge is an imprint of the Taylor & Francis Group, an informa business

British Library Cataloguing in Publication Data
A catalogue record for this book is available from the British Library

ISBN: 978-0-367-26640-0 (Set)
ISBN: 978-0-429-29433-4 (Set) (ebk)
ISBN: 978-0-367-27569-3 (Volume 1) (hbk)
ISBN: 978-0-367-27588-4 (Volume 1) (pbk)
ISBN: 978-0-429-29670-3 (Volume 1) (ebk)

Publisher's Note
The publisher has gone to great lengths to ensure the quality of this reprint but points out that some imperfections in the original copies may be apparent.

Disclaimer
The publisher has made every effort to trace copyright holders and would welcome correspondence from those they have been unable to trace.

POLITICS AND POVERTY

A critique of the Food and Agriculture
Organization of the United Nations

John Abbott

London and New York

First published 1992
by Routledge
11 New Fetter Lane, London EC4P 4EE

Simultaneously published in the USA and Canada
by Routledge
a division of Routledge, Chapman and Hall, Inc.
29 West 35th Street, New York, NY 10001

Printed and bound in Great Britain by
Biddles Ltd, Guildford and King's Lynn

British Library Cataloguing in Publication Data
Abbott, John *1919–*
    Politics and poverty : a critique of the Food and
    Agriculture Organization of the United Nations.
    1. Food and Agriculture Organization
    I. Title
    338.190601

    ISBN 0-415-06650-6

Library of Congress Cataloging in Publication Data
Abbott, John Cave, 1919–
    Politics and poverty : a critique of the Food and Agriculture
    Organization of the United Nations / John Abbott.
       p.   cm.
    Includes bibliographical references and index.
    ISBN 0-415-06650-6
    1. Food and Agriculture Organization of the United Nations.
2. Agricultural assistance. 3. Economic assistance. 4. World
politics—1945– 5. Poor—Developing countries. I. Title.
HD9000.1.F587A23 1992
630'.6'01—dc20
                                          90-26679
                                             CIP

CONTENTS

v

CONTENTS

PREFACE

FAO has been working on the food and agricultural problems of the world for over forty years. Why is there still such misery? Why does FAO have such a bad press? These are the questions that stimulated the writing of this book.

The FAO was founded by men of great ideals. The bulk of its staff joined it in the expectation that they would help to raise standards of living in countries where people were poor. Now, there are critics who damn international aid as wasteful, ineffective, even negative to welfare in the countries that receive it. Throughout the 1980s FAO's leadership has been perceived by many as self-perpetuating by whatever means, preoccupied with personal prestige, and egocentric towards authority.

How does the largest specialized UN agency operate in day-to-day practice? What has been the impact of Saouma, its Director General for the period 1976–93? How does he compare with his predecessors? What kind of staff has FAO attracted, and how well have they performed? My intention is to present an insider's view, to give a living dimension to what is often seen as an amorphous bureaucracy.

For some people, FAO in Rome is already equivocal. Northerners from America, Europe, and Japan may see the Mediterranean atmosphere as unconducive to hard work and productive discipline. Several times it has been suggested that FAO move away. Has its location in Rome been a handicap or an advantage? This is the matter of chapter 6.

Poverty will always be with us – in the sense that some people will always consider themselves to be less well off than others. Poverty at the level that people have not enough to eat need not be. It can be alleviated through increased food output by those of the poor who have access to land, and by expanded employment for those who have not. This depends much more than is generally realized on markets where farm output can be sold to pay for the seeds, fertilizers, and other inputs

ix

needed to produce more. The impact of drought and floods can be eased by the use of grain supply and price stabilization systems. In the words of a contemporary, 'Working on marketing and price policy in the Third World you may make it possible to increase the cash income of millions of peasants by a quarter or a third, and so save thousands of lives. You may be able to prevent a famine by a shift in price policy, or by an early warning system, as some of my colleagues have done.' In part II, the work of FAO is appraised in this strategic technical field. The constraints on Third World countries in the organization of agricultural marketing and input supply are set out, together with the assistance available through FAO, and the attitudes towards it of the governments concerned.

Pressure for the reform of FAO is mounting. The major contributors to its ongoing budget are tired of political resolutions that divert its conferences from useful discussion, they are tired of endless speechifying to give every member country a voice, and they are tired of governments that continue to absorb assistance but shy away from measures that could make it effective. Competition for use of the assistance funds still available from various sources is more intense. FAO must demonstrate the capacity to match this competition both in terms of economy and performance.

Rejuvenation of FAO calls for changes in a constitution designed for conditions quite different from those of today. It awaits measures to restore the pre-eminence of its technical as opposed to its representational departments, to promote more productive leadership, and to favour the channelling of aid resources through it as against a wasteful dispersal. One of these measures impinges on the most sacred of UN cows – the importance of one country, one vote, irrespective of size and validity of country.

The assistance, albeit unwitting, of a host of FAO colleagues and friends is gratefully acknowledged. Substantial use has been made of the *CASA Gazette*, an informal publication of the FAO staff co-operative. Thanks are expressed to all those who contributed to its pages, including K.A. Bennett and John Spears, whose writings have been reproduced. Above all, appreciation is expressed of FAO for the use made of its documents, the challenges it offered, and for the rich experience obtained by working with it. The author concludes with the conventional disclaimer – he alone is responsible for the statements made and the opinions expressed.

ABBREVIATIONS

AGRIS	International Information System for the Agricultural Sciences and Technology
CFA	Committee on Food Aid Policies and Programmes
CGIAR	Consultative Group on International Agricultural Research
CIMMYT	International Maize and Wheat Improvement Centre
FAO	Food and Agriculture Organization of the United Nations
FFHC	Freedom from Hunger Campaign
IFAD	International Fund for Agricultural Development
IFPRI	International Food Policy Research Institute
IIA	International Institute of Agriculture
ILO	International Labour Organization
IRRI	International Rice Research Institute
OPEC	Organization of Petroleum Exporting Countries
TCDC	Technical Co-operation among Developing Countries
TCP	Technical Co-operation Programme
UN	United Nations
UNCTAD	United Nations Conference on Trade and Development
UNDP	United Nations Development Programme
UNIDO	United Nations Industrial Development Organization
UNRRA	United Nations Relief and Rehabilitation Administration
UNSF	United Nations Special Fund
USDA	United States Department of Agriculture
USAID	United States Administration for International Development
WFC	World Food Council
WFP	World Food Programme

1

INTRODUCTION
Storm clouds mar the ideal

In 1935 Frank McDougall, economic adviser to the Australian mission to the League of Nations, cabled the nutritionist Boyd Orr in Aberdeen: 'We have today lighted such a candle, by God's grace, in Geneva as we trust shall never be put out'. He was echoing the words of the religious martyr Hugh Latimer to his fellow protestant Ridley in 1555 as they were about to be burned at the stake.

Stanley Bruce, High Commissioner in London and former Prime Minister of Australia, reacted against the prevailing view in the 1930s that the current surpluses of agricultural commodities could only be handled by cutting down production. Together Bruce, Boyd Orr, and McDougall had worked on 'a marriage of health and agriculture'. McDougall had written a memorandum, 'The Agricultural and Health Problems'. He said it would 'argue a bankruptcy of statesmanship if it should prove impossible to bring together a great unsatisfied need for highly nutritious food and the immense potential production of modern agriculture'.

Bruce picked this up in an address to the League of Nations. The favourable reaction it received inspired McDougall's famous telegram. Such was the feeling of the men who laid the basis for the establishment of FAO.

A MARRIAGE OF HEALTH AND AGRICULTURE

There had already been another man convinced from his own experience of the need to help the small-scale farmers of the world, who also had the vision and determination to put his ideas into effect. This man was David Lubin. He emigrated from Poland to the USA in the nineteenth century. There he farmed and became an agricultural merchant in California. The difficult working conditions experienced by

1

farmers there during the depressions of 1880–90 made such an impression on him that he tried to set up some international mechanism to better their lot. This crusade led him to Italy, where he impressed King Victor Emmanuel III. In 1905 the King communicated to his Prime Minister that 'it would be extremely useful to set up an international institute which, without any political design, would study the conditions of agriculture in the various countries of the world and periodically issue information on the quantity and quality of crops.' This led to a conference in Rome and the signing by forty countries of a 'Convention Establishing an International Institute of Agriculture'. It began its work in the Villa Borghese, Rome.

The Institute collected statistics, carried out studies, and gave out information. Bruce, McDougall, and Boyd Orr – all Scots by origin – went much further. They wanted to set up a body that would manage the world's food resources, using its surpluses to the best advantage. However, influential world opinion still had to be convinced of the vital links between nutrition, agriculture, and economic development.

McDougall went to Washington in 1941 and 1942, representing Australia in negotiations for an international wheat agreement. Together with a group in the United States Department of Agriculture he wrote a further memorandum on 'A United Nations Programme for Freedom from Want of Food'. This set out ideas on how governments might develop an organization to deal with food and agricultural problems. His next great moment came when Mrs Eleanor Roosevelt heard of his memorandum, met McDougall, and decided that his ideas were worth bringing to President Roosevelt's attention. McDougall was invited to dinner. He seized the opportunity to outline his concept for an international agency concerned with food and agriculture. Thereafter came a long phase of deflation – McDougall heard no more. Only from a newspaper did he learn that in September 1943 President Roosevelt was going to invite the allied governments to hold their first conference on food and agriculture at Hot Springs, Virginia. This conference set up an interim commission which led to the establishment of FAO at a conference in Quebec on 16 October 1945. McDougall was a member of the Australian delegation to the conference, and he was appointed Counsellor to the FAO Director General – a role in which he continued until his death. His grave is well marked in the Protestant Cemetery in Rome.

At each biennial FAO governing conference an eminent statesman is invited to give the McDougall lecture, a tradition established in his honour.

FREEDOM FROM HUNGER

A second phase of exaltation for FAO came in the 1960s with the election of Director General B.R. Sen. In addition to obtaining major increases in its budget (voted directly by member countries), FAO was assigned a leading role in the use of the greatly increased resources pledged to a UN Special Fund for Development. This enabled FAO to mount influential teams of technical advisers, and to organize large-scale training for its developing member countries, at a time when this assistance was highly appreciated. The bulk of its Third World members had just achieved their independence. They looked to FAO for an international frame of assistance in contrast to the single voice of their former colonial power.

In his autobiography (Sen 1982), Sen averred that his greatest satisfaction was the successful launching of the FAO Freedom from Hunger Campaign which established FAO as a vehicle for the use of aid funds contributed voluntarily by religious and other voluntary groups in North America, Europe, and Australasia. The funds they collected would support development projects that FAO would implement on their behalf. The 1960 conference of international non-governmental organizations gave FAO a new dimension as the world focus of voluntary development efforts in its field. A special assembly of eminent personalities in Rome in March 1963 issued their manifesto, 'Man's Right to Freedom from Hunger'.

A World Food Congress was held in Washington in June 1963. It was here that Sen also gained the support of developed countries' industry concerned with food and agriculture. The FFHC Fertilizer Programme brought resources in kind and cash to FAO to implement field trials and demonstrations of fertilizer use. Food processing industries also offered collaboration in identifying and implementing practical development projects, lending experienced personnel, etc.

The FAO Indicative World Plan also stems from this Congress. The world wanted some indication of the magnitude of the food and food production resources that would be needed to match the expected growth in world population. The FAO projected figures for 1985 were the first ones developed in detail for a world development frame.

At this time FAO could also face up directly to the population control issue and call upon the Pope to help. To the Vatican argument that resources will always be found to meet the food needs of larger populations, Sen responded that they are not in practice available at particular places and times; poverty and famine can be the result.

PROVISION OF FOOD AND CAPITAL

Further great days for FAO saw the establishment of the UN/FAO World Food Programme and of the FAO Cooperative Programme with the World Bank. Back in the 1940s Boyd Orr, the first Director General of FAO, had foreseen these needs. He wanted the FAO itself to be able to supply food where people needed it in order to survive. He wanted to be able to finance the provision of infrastructure, equipment, fertilizers, and other inputs for expanded and more productive agricultural and fishery operations. At the time his ideas were thought premature. In 1961 the World Food Programme was established in Rome as a joint arm of FAO and the United Nations. Its goals were to make surplus food available to avert famine, and also to pay for labour on projects to improve rural infrastructure and on other development activities.

A.H. Boerma became the first executive director of the World Food Programme. His experience at the end of World War II trying to organize food supplies for a poverty-stricken Netherlands had made him an advocate of a 'world food bank'. His elation at this appointment was great indeed.

Shortly thereafter Sen made his historic agreement with George Woods, President of the World Bank. Up to this point its level of lending for agriculture in the developing world had been limited. Woods, however, saw the need for a greatly expanded flow of capital into Third World agriculture. On hearing this, Sen recalls, 'I took a plane to New York to meet him.' The outcome was an arrangement whereby FAO would identify and prepare potential projects for financing. The World Bank would appraise and supervise them. It undertook to contribute two-thirds of the cost of a team of specialists located in FAO that would concentrate on agricultural and fisheries work for the World Bank. With the pledging of funds for a parallel International Development Association programme to serve countries unable to meet commercial interest rates, this opened the way to a massive channelling of new capital into Third World agriculture. The array of development instruments was now complete, with FAO at the centre.

DEDICATED STAFF AND VOLUNTEERS

Morale was high in FAO throughout the 1960s – at all levels. There was strong competition for all posts as they became vacant. From the UK, especially, young people joined FAO at levels far below their qualifications because they saw its work as exciting, with good

4

prospects; they were pleased to talk about FAO and to be part of it.

My recollection of my own feelings on joining FAO is still vivid. I first heard about the Organization through a talk at my university by Boerma, then its Director of Economics. He set out its scope of work and objectives. Then he appealed to young men and women to give up narrowly circumscribed, over-staffed fields of work in the libraries and laboratories of industrial countries' universities, and take up instead the challenge of helping the developing world. I was hooked. At the reception which followed he said there would be work in FAO in my field and encouraged me to apply for a post. When the reply to my application came – positive – a great warmth surged through me. My wife saw it and marched off into the woods – she knew she had lost me.

I then tried to inform myself about FAO and its place of work. The first indications were none too encouraging. My only contact was a graduate student friend who had done a year at FAO and then left. He did not like it. In conversation, he said that the only good thing he saw in FAO was that it was an easy drive to the beach; one could get there conveniently for a swim after work. He went on to become Director of Agriculture and Rural Development at the World Bank. I had also heard tell how an FAO representative had parked his car outside a hotel in Rome for an official luncheon. When he came out, it was set up on bricks – minus its wheels. I talked in a bar in Sacramento to a waiter who came from Rome. 'What kind of a place is it now?' I asked. 'There are some good night clubs there', he said, 'but not much else'.

My old professor at Oxford also held a negative attitude toward FAO. 'They will take you for a year or two, draw out any ideas you

Figure 1 The FAO emblem

have, then drop you like a squeezed-out lemon,' he said. I later found the squeezing could be mutual and productive. FAO, it is true, wanted my ideas for a manual on agricultural marketing improvement in the developing countries. At the same time, it put so much information at my disposal, and such rich opportunities to gather experience, that I felt myself steadily enriched by the exchange.

My work was tremendously exciting. It gave me a chance to exercise in combination my professional training, my practical management talents, and my engrossing interest in the developing world. People throughout the developing countries appreciated the seminars and training programmes we organized. In seven or eight languages they read our books at the policy-making level, in colleges and for use in extension work. The books were cheaply priced and sold by the tens of thousands. Many who read them were from countries where we helped to develop their own services in our technical field, helped strengthen their agricultural economies, provided a more satisfying range of foods to consumers, and put money in the pockets of their farmers. We were also adept at making the most use of the money that was provided; which in itself was stretched by collaboration with bilateral agencies and other UN programmes. Projects integrated into our overall programme were formulated in advance, awaiting funds that might be released towards the end of a budget period. FAO went a long way on a little money at that time.

With technical committees examining the work of FAO in some detail there was a solid professional interest in the biennial FAO conference. We hoped it could be attended by people who knew our subject so that there could be some constructive comment.

Volunteers of all kinds came to FAO offering their services during its first decades. Most numerous were young people from the universities, professionally qualified, and seeking direct experience in the developing world. As associate experts they were incorporated into FAO advisory teams, their expenses covered by allowances from their governments, but with no salary. There were older men, already well-off, who worked for FAO for a nominal one dollar per year. There was the American professor who heard of our idea of establishing a listing with summary notes of materials that would be useful in briefing field advisers in marketing. This needed some extra money for correspondents in various countries, and for typing services. He found it for us from a private foundation.

NIGGLING AT THE EDGES

FAO had its problems and faced some attacks during its first decades, but they were minor. The English newspaper the *Daily Express* ran a series of articles in the 1950s and 1960s criticizing the income tax and commissary privileges of FAO staff. Stirring up popular jealousy was a good journalistic line in then-socialist Britain, with its low salaries and punitive taxation.

There was also in-fighting with UNDP. FAO had its own representatives in a few countries. Some maintained a pugnacious independence. DG Sen put them firmly into UNDP country offices as senior agricultural advisers. The establishment of the UN Special Fund in the 1960s had brought greatly increased resources to FAO as its main agency of implementation. For 1975 FAO was allocated $122 million in aid funds, one-third of UNDP's total resources.

When Director General Saouma established separate FAO representatives with their own direct link to the country's government, UNDP turned away from FAO. Projects on the margin of FAO's competence were assigned to other agencies; FAO was considered 'too big'. The 1990 UN proposal that its aid funds should go directly to recipient governments would cut FAO down substantially. Governments might engage FAO assistance with part of the funds provided: they might also be besieged by consulting firms seeking direct contracts and offering inducements to obtain them. In June 1990 FAO estimated it would lose 450 experts and headquarters support posts.

Periodic meetings of government representatives with FAO specialists and consultants were held by its commodities division to discuss prices and prospects for major products in international trade. These commodity study groups tended to grow up and leave the brood. Independent organizations such as the International Wheat Council took over their role. The tendency for these commodity bodies to be influenced by the major exporters then sparked the establishment of the UN Trade and Development Conference to press the Third World's interest. Taking a line of confrontation, UNCTAD's leader, Prebisch, also wanted his own commodity analyses, duplicating those of FAO.

The International Trade Centre grew out of GATT (The General Agreement on Trade and Tariffs) to help the developing countries find export markets for their products. It took over work in the agricultural area that had formerly been done by FAO, albeit with limited resources. With UNESCO and UNIDO, FAO had boundary disputes over agricultural education and training, and agricultural processing respectively.

At the Fourth World Food Conference held in Rome in 1974, the process of setting up new organizations to take over elements of FAO's work gained new momentum. The International Food Policy Research Institute in Washington was given funds to carry out studies fully within the scope of FAO's economic and social department. While FAO already had a small unit with a mandate to co-ordinate agricultural research, a new international institute with a professional staff of fifty-five was set up to this end in the Netherlands. FAO's original mandate continued, but these new organizations could apply to some areas of its work much greater resources.

The conference of 1974 also established a World Food Council. FAO's biennial conference was attended, at best, by ministers of agriculture (and of food, if this was combined in the same ministry). The World Food Council, to be attended by ministers of foreign affairs, was intended to secure greater government commitment to food security. It was located in Rome, but was not part of FAO. It could be seen as backing up FAO's attempts to achieve its programme objectives. It could also be seen to be checking that FAO was doing its job.

The attack on the transnational companies mounted in the early 1970s was also negative for FAO. Because FAO had arrangements for collaboration over fertilizer promotion, the use of agricultural machinery, and the adaptation of processing technologies to Third World conditions, it was criticized as being in their hands. There were transnational excesses – for instance the ITT political involvement in Chile, and the promotion of baby food by Nestlé in areas where incomes were very low and mother's milk was a better recourse. Small countries were afraid of the transnationals because of their financial clout. However, transnationals with immense resources, able to command the best brains and with an incentive for performance rarely found in public enterprises, are a fact of life. Failure to take advantage of the products of their research and development and of their managerial know-how in implementation has its cost. FAO shed the Industry Co-operative Programme, but maintained its close links with the European fertilizer associations.

UN POSTURING AND AID FATIGUE

FAO suffered, along with other UN agencies, from the inclination of its developing country members to bring up 'political' resolutions that were aside from its programme of work. A regional conference in Tunis attended by delegates from many countries and numerous FAO staff had to be halted midway on this account. Generally, western donor countries listened in silence and abstained. The FAO's insistence on one such resolution at the 1989 FAO Conference, however, provoked the USA to threaten withdrawal. The USA and the UK had already withdrawn from UNESCO in protest at its management and programmes.

With the advent of Thatcher–Reaganism, attitudes in the UK and USA towards the United Nations cooled. The United States in particular reacted against the voting of large budgets for UN agencies by a majority made up of small countries that contributed little themselves. Congress passed a law limiting the USA's contribution to a 15 per cent share where voting power was not weighted by financial contribution. For many years the USA had contributed 25 per cent of the FAO budget. The payment of the US contribution was also delayed.

There was also a growing scepticism as to the effectiveness of international aid. Well known in Britain from the 1950s were three economists of Hungarian origin – Balogh, Kaldor, and Bauer. The first two became socialist establishment figures in the 1960s and were then raised to the peerage. Bauer remained in dissent. 'If a country cannot progress without aid,' he maintained, 'it certainly will not do so with it.' He saw no grounds for the guilt richer countries felt, which stimulated them to provide aid which he felt would, in any case, not be effective. 'Too polemic to gain acceptance' was the academic judgment on his writings. They kept appearing, however, in books with provocative titles, and as articles in the conservative press. They were also taken up by a minority of Third World students who felt proud of their independence of thought.

With Margaret Thatcher's election as the UK's Prime Minister Bauer's turn came. Through her he became Lord Bauer. How well the bestowal of this honour reflected the shift of political opinion. Throughout the 1960s and 1970s Britain and the USA were under constant internal criticism for reserving part of their aid for the use of their *own* expertise and equipment. For aid to be 'good' it should be free of such ties. These countries were also seen as laggards in attaining the UN aid target of 1 per cent of GNP (the value of a country's total output). In 1985, the Minister of State Timothy Raison could maintain in the UK House of Commons that it was right 'that the bulk of our bilateral aid should be tied to expenditure on British manpower and materials'. Total aid budgets were cut in the UK and USA along with other government expenditures.

Total pledging of funds to the UN Development Programme has

been reduced sharply under the same aura of malaise in the UN system. In 1989 only Denmark, Norway, and Sweden assigned 1 per cent of their GNP to international aid; the average for the developed countries was 0.33 per cent. In the USA this was attributed to 'aid fatigue'. Many of the countries seeking aid, it was said, were led by governments dictatorial in their procedures, tolerant of corruption, and committed to policies deemed counter productive. After decades of assistance, most of Africa, in particular, had receded rather than advanced. Was it useful to give it more aid?

TEMPERAMENT DISTORTS THE IMAGE

An intrinsic feature of FAO is that it is directly responsible to its member countries. In consequence it has always been wary of offending them. That a representative could stand up at a biennial conference and criticize the performance of an FAO officer in relation to his country was healthy – there was an incentive to keep the decks clear. However, there were disadvantages too. The original figures for Latin America in the Indicative World Plan could not be issued: they were too pessimistic (i.e. too realistic). Analytical documents lost their bite – criticisms of particular governments had been wrapped up so elaborately that they were detectable only by those already in the know. FAO statistics are widely used throughout the world. They are 'official', which means that they conform to the figures issued by governments, even though it is known that these may not in themselves be correct. This possibility is, however, taken into account in analytical studies using such figures.

Under Director General Saouma, however, the need to avoid offending individual governments was paramount. The funds available for direct assistance under the Food Loss Prevention and Technical Co-operation Programmes were pre-allocated country by country under his immediate supervision. A project proposal considered unsound by a technical unit could still be approved by Saouma.

The inclusion of some twenty-four African countries in a 1984 list of those in danger of famine and needing assistance was blamed for distracting attention away from the much more serious plight of Ethiopia (Gill 1986). However, in various countries, since the 1970 drought in the Sahel, food aid had become an established resource. Government agencies were financed from the proceeds of its sale; FAO under Saouma could not risk offending any one of them by omitting it from the list.

Delays of three months or more in the issue of FAO/WFP reports on the potential famine situation in Ethiopia in 1983 were also criticized. FAO has always been wary of publicizing a need for assistance without a government's prior agreement. Such a report from FAO – likely to be treated as authoritative by governments world-wide, by non-government aid agencies, and by journalists looking for a new sensation – could backfire on the government concerned. Its task of relieving incipient famine could be complicated both by hoarding on the part of those in possession of supplies, and alarm among those without them.

WFP shipments of emergency food aid require the approval of the Director General of FAO. He was further criticized in June 1984 for delaying the despatch of a 25,000-ton consignment to Ethiopia. The Canadian television film prepared for the 1987 election claimed that Saouma withheld authorization for twenty days until an Ethiopian representative in Rome that he disliked was removed, and that 30,000 people died of famine in the period corresponding to this delay. Saouma denied this.

The most consistent critic of FAO's management by Saouma was Otto Matzke. He maintained a continuing flow of articles in the *Neue Zürcher Zeitung* pointing to distortions and corruption in its direction. Matzke was a Director in the World Food Programme under Boerma. He also had excellent sources of information on FAO. At the FAO meeting of senior staff which honoured his retirement, he claimed that he wrote with Boerma's concurrence; criticism, it had been agreed, was healthy.

Saouma provided a more generous target. Major points of attack for Matzke were:

1   FAO was far too large. With 7,000 (10,000, he claimed) professionals, general service and consultant staff in Rome, country and regional offices, and field posts, there could be no adequate supervision. Too many were there 'to meet programme delivery requirements' without due provision for their efficiency and development impact.

2   The establishment and staffing of FAO representative offices in a large number of countries, in addition to the pre-existing regional structure, was a waste of public funds. Prime considerations, he maintained, were its use for patronage appointments and the promotion of Saouma's re-election as Director General. He also attacked Saouma's Technical Co-operation Programme (funding from the FAO budget of direct assistance to a member country at the director general's discretion) alleging its misuse as a re-election mechanism.

3 Insufficient co-ordination with other UN agencies, especially UNDP.
4 Political use of emergency food aid.

Matzke was right on all these counts. While the gist of his criticism also appeared in the international press, his main audience was German-speaking. For more than a year there was a direct stand-off with the West German government over a project which FAO had implemented in the Sahel with West German funds. The West German government wished to check the use of its funds; Saouma would not agree. He would accept no other auditing than FAO's own. The culmination of Matzke's criticisms of Saouma's management of FAO was a full-scale debate in the West German parliament. There were several Germans in high posts in FAO. These were mobilized in the Director General's support. Saouma also cultivated his relationship with the minister responsible for aid. Despite the adverse publicity, West Germany still voted for Saouma at the 1987 election. Some donor countries, however, have cut down the number of projects they entrust to FAO for implementation.

Some of the clouds over FAO are shared with the whole UN system. When somebody else pays it is easy to approach each new problem with the same non-solution – set up another UN agency. Further UN machinery will then be needed for co-ordination. More well-paid posts will then be open for those who backed each proposal and prepared the documents. Tired of expanding bureaucracies and agencies on whose policies they can have little influence, the reaction of those who provide the money will be to starve them of funds. The growth of the CGIAR (Consultative Group on International Agricultural Research) chain of research institutes with major donor participation in their management is a response to this process of events. Certainly, if FAO is to thrive again it must be freed of the influences so prejudicial to its professional morale. It can also be much smaller than it has been, cost less, and be more effective. Proposals for its reform are outlined in the final chapter of this book. That there will be a continuing need for a world organization equipped to provide professional guidance in food and agricultural development is clear; these two fields are still vital factors in so many countries. A realistic view of many African nations is that not twenty but one hundred years will pass before they have the internal strength to face the world alone (Cockroft 1990).

Funds for development aid will still be forthcoming – if not through one channel, then through another. Those responsible for implementing bilateral and non-government aid projects know well how weak their position is without experienced technical support. It is up to the management of FAO to demonstrate what it can offer to the EEC, Japan, and other future giants in aid potential.

Part I

THE ORGANIZATION AND ITS STAFF

2

THE ORGANIZATION

FAO first took shape at the conference called by President Roosevelt at Hot Springs, Virginia, 18 May to 3 June 1943. Forty-four countries participated. They issued a declaration of intent, and an interim commission was appointed to draft a constitution and set out what the new organization should do. FAO was established officially at the first session of its governing conference in Quebec in 1945. At this time the United Nations as a body was yet to be formed. The name agreed upon, 'The Food and Agriculture Organization of the United Nations', was intended to emphasize its independent status and at the same time relate it to the UN which was soon to be established. It was voted an annual budget of five million dollars. The USA undertook to contribute 25 per cent, the UK 15 per cent.

MEMBER COUNTRIES

FAO is run by its member countries. By 1990 they included most independent countries with the notable exception of the USSR. Their numbers have increased steadily as new, and smaller, countries attained independence, reaching 158 in the mid-1980s. With membership comes the obligation to contribute towards FAO's operating budget in proportion to a country's national income. For developing countries this is not onerous; they benefit from FAO's technical assistance and tend to apply for membership as soon as they become independent. A few of them, however, (Cameroon, Ivory Coast, Malawi, for example) were cautious in seeking FAO assistance, preferring to hold on to that of their former metropolitan country.

The USSR attended the Quebec Conference along with the other countries on the winning side in World War II. At a late stage its delegates received instructions from Moscow to be observers only. The official Soviet position was that the 'imperialist powers' would fill the key posts and use the organization to their advantage. The other 'communist' countries in FAO's original membership left in the 1950s but have since re-joined. The USSR has stayed aloof. Views on the reasons for this are wide-ranging – the relatively poor performance of its agriculture meant that it had nothing to show the developing countries; participation in FAO could mean that its official statistics would come under scrutiny; and, according to Sen, there was a simple unwillingness to fork out a contribution in convertible currency. There were hopes that the USSR might come in at the time of the 1974 World Food Conference. It was entitled to a seat there as a member of the United Nations. R. Aubrac, an FAO official known for his communist contacts, was sent to Moscow to urge participation. He came back with a statement by a minister, 'There is an empty seat that has to be filled.' It is still vacant sixteen years later.

The USSR's absence from the organization did not mean that FAO took no notice of Russia's very considerable agricultural economy. Russian-origin FAO staff continually monitored its publications and press reports. Its impact on world food supplies and agricultural trade was studied. Also, large amounts of roubles came to FAO through the UN for use in technical assistance. They were not convertible, so their use became an exercise in ingenuity. FAO seminars were organized in Russia, focusing on aspects of agriculture in which it was advanced – irrigation, dry land farming, etc. Nylon fishing nets were bought there for demonstration use in developing countries. At one time, FAO bought Russian jeeps for its field staff – until delays over spare parts brought the programme to a halt.

Under Gorbachev and *glasnost* the USSR is expected to join FAO very shortly. In May 1989 *Moscow News* featured an article asking why the USSR remained outside FAO. It set out potential advantages to the USSR in joining the organization, which would be a source of useful information and experience and a forum for discussing important international issues, such as the protection of the environment. Part of the annual $28–30 million membership 'fee' would come back in salaries to some seventy Russian professionals that would be employed by FAO. Perhaps some of the balance could be paid in kind?

China's re-joining of FAO in 1973 was a major event for the organization. There were the debts of the old Chiang Kai-shek government to be written off, the word Taiwan to be eliminated from documents, a book written by a member of the FAO Far East regional staff to be pulped, Chinese to be taken up again as an official language.

The Chinese government provided a team of translators and interpreters, meeting part of the cost. FAO had to obtain Chinese language typewriters. The Chief of Conference Services got into trouble over a set he bought for them in Hong Kong. These machines came with up to 600 Chinese characters, but not in the combination or form needed by the team from the People's Republic. The Chinese team kept very much to themselves. They wore a grey uniform, arriving and leaving together in their own minibus. I asked one if she was learning Italian; she said no. I asked 'Don't you need it to go shopping?' 'We don't go to shops,' was the reply, 'all our requirements are met by the Embassy.' Through the 1970s the People's Republic was proud of its achievements founded on human effort and self-reliance. They were also fashionable with intellectuals in the West. Briefing tours to China were arranged for FAO staff. It later came out that many deficiencies in the Chinese agricultural economy had been hidden. China became a major user of FAO's technical assistance.

THE WORK OF FAO

The organization was set up to:

- raise the level of nutrition and of living of the people in its member countries;
- improve the production and distribution of food and agricultural products;
- improve the condition of rural populations;
- contribute towards an expanding world economy and ensure humanity's freedom from hunger.

It would also constitute a means through which member countries would report to each other on measures taken and progress achieved.

The FAO constitution enlarges on how it should achieve its objectives, indicating that it should undertake the following.

1   Collect, analyse, interpret and disseminate information on nutrition, food and agriculture, including fishing and forestry.
2   Promote and recommend national and international action for:
    - research and the improvement of education and administration relating to nutrition, food and agriculture;
    - the conservation of natural resources and adoption of improved methods of production;
    - the improvement of processing, marketing and distribution;

- the provision of adequate national and international agricultural credit;
- the adoption of international policies on agricultural commodity arrangements.

3 Furnish such technical assistance as governments may request, organize missions to help them meet the recommendations of the conference setting up FAO and those of its constitution, and take all necessary and appropriate action to implement its purposes.

To carry out these objectives, FAO proposed to its member countries for the years 1988 and 1989 an annual budget of $225 million. It expected to receive a further $700 million to finance its technical assistance to individual countries. This would come from UNDP in New York, to which funds are pledged by UN member countries, and from individual government, religious, and other aid agencies which use FAO to manage projects for them.

Of the $225 million in FAO's own budget, some $30 million would also be used on direct assistance. Since 1978 FAO has had its own Technical Cooperation Programme (TCP). This provides technical and material assistance at short notice up to a maximum of $400,000 annually per country. This can be strategically important. The rather lengthy procedures for funding through UNDP and some bilateral sources can result in men and equipment arriving after the opportune time has passed.

The balance of FAO's budget, plus a 13 to 14 per cent commission on field projects funded from other sources, goes to support its technical, administration and representative services at headquarters, in the regions and in individual countries.

ORGANIZATIONAL STRUCTURE

The governing bodies of FAO are its Conference of member governments which meets every second year and its Council of forty-nine government representatives selected regionally. The council meets twice a year. Its Finance and Programme Committees have a role in examining FAO performance and proposals in some detail, as do Committees on Agriculture, Fisheries and Forestry. The council itself is now sadly deteriorated. Once it was seen as 'The World Food Council' where eminent representatives debated and advised on major issues.

Draft programmes of work, biennial budgets, and proposals for changes are generally prepared by FAO staff and presented to these bodies for comment and approval. The changes they make are necessarily relatively minor, due to the participants' lack of knowledge of the detailed working of FAO. Sessions of the FAO conference and council can be very boring, with ministers or their deputies presenting a succession of carefully tailored country statements. It can liven up on occasion, however, as in October 1988 when Bula Hoyos (Colombian), one-time chairman of the council and candidate for Director General, called the American ambassador 'a fat, bald slob'.

A simplified organizational structure of FAO is presented in figure 2. Included with the Director General are his cabinet, programme, budget, and evaluation; inter-agency affairs, the legal office and internal audit are also in his office.

There are seven departments. The divisions listed under each give a broad indication of their work areas and responsibilities. The Agricultural Services Division (AGS) deals with the main commercial services to agriculture, i.e. farm management, agricultural engineering and storage, processing, marketing, provision of seed and fertilizers, etc., and credit. The Research and Technology Division promotes and co-ordinates national research efforts in agriculture. It does not undertake its own research; this is not in FAO's mandate. The units headed Operations manage FAO's field projects in collaboration with the relevant technical groups.

The Development Department is responsible for the FAO country representatives and the preparation or review of technical assistance or investment projects. These may then be handled under FAO's Technical Co-operation Programme or be taken up by international financing agencies like the World Bank, or financed by bilateral aid. Projects are also offered to non-governmental donor groups participating in the Freedom from Hunger Campaign, which was instituted by Director General Sen in the 1960s to provide a convenient channel for voluntary assistance.

The Administration and Finance Department manages the finances of the FAO, recruits and terminates staff, determines their allowances, and pays them. Not least, it maintains the buildings of FAO, its office furniture and equipment, including increasingly important computer services.

Provision of information, international documentation services, the library, and the servicing of the conference, council and technical meetings are the main responsibilities of the Department of General Affairs and Information.

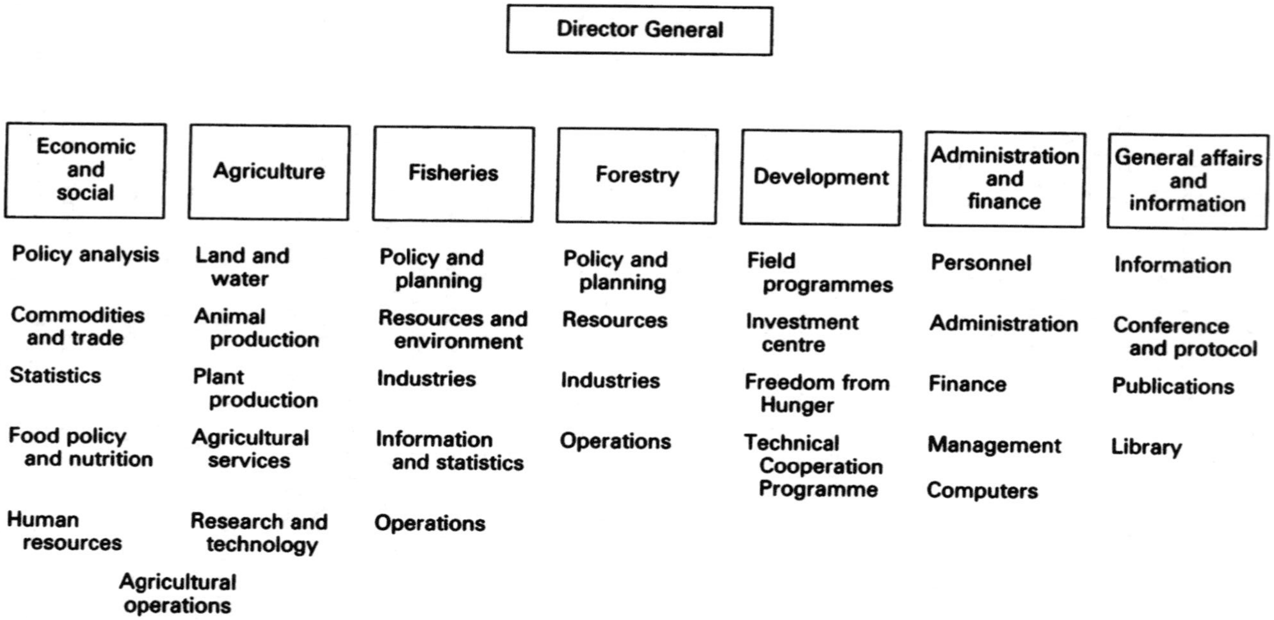

Director General

Economic and social	Agriculture	Fisheries	Forestry	Development	Administration and finance	General affairs and information
Policy analysis	Land and water	Policy and planning	Policy and planning	Field programmes	Personnel	Information
Commodities and trade	Animal production	Resources and environment	Resources	Investment centre	Administration	Conference and protocol
Statistics	Plant production	Industries	Industries	Freedom from Hunger	Finance	Publications
Food policy and nutrition	Agricultural services	Information and statistics	Operations	Technical Cooperation Programme	Management	Library
Human resources	Research and technology	Operations			Computers	
	Agricultural operations					

Figure 2 FAO organizational structure

The total staff of FAO at its Rome headquarters is around 4,200 – if all the posts are occupied. With some government contributions withheld or delayed in the late 1980s many posts had to be frozen when they became vacant. Of the established posts some 1,600 are professional or above, and 2,600 are general service posts.

Regional offices

FAO has a regional office headed by an Assistant Director General, with a team of technical staff, in Accra, Bangkok, and Santiago. They serve its African, Asian and Pacific, and Latin American regions. Its office in Cairo for the Near East and North Africa was closed when most Arab countries broke with Egypt over the Camp David agreement with Israel. It is scheduled to re-open soon. There are also liaison units in Rome for Europe, in Washington for North America, in New York for the United Nations, and there are staff assigned to agriculture divisions maintained jointly with the United Nations Regional Economic Commissions in Addis Ababa, Baghdad, Santiago, and Geneva. In 1981 some 130 professional or higher category officers, and 215 general service staff, were budgeted for these regional offices, liaison units, and joint divisions. In 1989 the Bangkok office housed thirty professionals, with supporting staff of 105.

These offices were established to bring FAO nearer to its member countries. They organize conferences every other year to enable governments of the region to report on and discuss food and agricultural issues. The technical staff organize assistance programmes on a regional basis to match common language needs and common economic and cultural backgrounds. These considerations have often proved illusory. Bangkok has remained a very popular centre for regional activities with transport convenience, low cost of living, and other recognized attractions. The other regional headquarters have had their difficulties. For example, the representative for the Near East had to be transferred to Rome for a number of years when Egypt made peace with Israel. Santiago has been popular with some people at one time and with others at another. It has never been a convenient base for activities ranging to Mexico and the Caribbean, over countries with English, French, and Portuguese national languages as well as Spanish. A practical advantage to officers posted there was that for many years they could import Mercedes cars and after four years sell them for double the original price.

Certainly it was a mistake to establish an FAO regional centre for

Africa in Accra. There were arguments for it at the time. Ghana was the first long-time colony in Africa to achieve its independence. Its leader Nkrumah was the speaker for new Africa. He was so popular that in the words of an FAO staff member, 'hundreds of women gathered at the door of the conference hall and when he arrived threw down their robes for him to walk on'. Accra contained many well-educated people. It was reasonably well located. However, the United Nations Economic Commission for Africa and the Organization of African Unity went to Addis Ababa, capital of the oldest independent African country. Nkrumah began a squandering of Ghana's resources that resulted in FAO regional staff travelling weekly to Lomé in Togo to obtain basic necessities. No other agency came to join FAO in Accra; well qualified people would not work there.

Country representatives

These were established systematically by Director General Saouma in his 1976 and subsequent programmes of work to a total of seventy-two. The concept, however, was by no means new. Wherever FAO had a substantial number of officers working there was a need for a leader to co-ordinate their position on matters of government support, to help newcomers make contacts and, where necessary, to reinforce a professional position. Commonly the most senior, or the one whose terms of reference gave him an appropriate mandate (e.g. a planning adviser), would be appointed 'country representative'. This appointment could be a kiss of death. Specialists lost momentum in their technical field. There were a few who became 'professional representatives', provided with an office by the ministry of agriculture. They would then engage in petty in-fighting for precedence with the resident United Nations representative. It was to avoid this and to take advantage of a UN offer to fund two-thirds of their cost that DG Sen agreed to a system of senior agricultural advisers/FAO country representatives located in offices provided by UNDP. At that time it was UNDP, essentially, that financed the FAO field programmes.

DG Saouma's grounds for taking over to FAO the entire cost of a representative so that he could operate them independently was that he foresaw considerable funding from other sources. This was initially expected from the Arab governments that sponsored him; however, their support never became substantial. It was forthcoming instead from northern European countries pursuing the aid target of one per cent of their gross national product. In countries such as Peru and Tanzania bilateral aid channelled through FAO eventually became 60 per cent of its total funding and the place for an FAO representative became clear. In addition to helping identify and follow up projects for external assistance, the FAO representative channels information between FAO and the government, seeks participation in FAO global programmes, including celebration of World Food Day on 16 October, and facilitates visits by FAO personnel. With one senior officer, a programme assistant, a secretary, and a car, in an office provided by the government, the cost to FAO was around $150,000 per year.

Field projects

If there is a justification for FAO's existence on its present scale it is its ability to put into a developing country people who can demonstrate the advantages of an irrigation pump, train local people to use it, show farmers how to obtain higher yields from their crops, help them get better prices for their output, and build on sound foundations an economic and social infrastructure for the future. FAO provides technical assistance to its developing member countries over its whole subject coverage – agriculture, fisheries, forestry and nutrition. It helps governments to strengthen their economies, improve food supplies for their people, and develop rural industries. The training of their own nationals both by working alongside international experts and by participating in specific programmes is a major feature of FAO work.

In the early 1980s there were about 1,500 field officers on fairly long assignments, 600 short-term consultants, and over 200 associate experts (financed by developed country governments to acquire practical experience) working on the economic and technical aspects of agriculture. These amounted to about 60 per cent of FAO's field staff with a further 40 per cent working on fisheries and forest improvement. These men and women were spread over 130 countries. The largest numbers – say over 100 – were working in countries such as Mozambique and Tanzania, whose governments were looking specifically to the UN agencies to help them through difficult phases of rehabilitation and development. The cost of running this programme of assistance to agriculture was around $160 million per year. At that time, two-thirds of the money came from UNDP, over twenty per cent from funds placed in trust with FAO from bilateral and other external sources, and rather less than ten per cent from FAO's own Technical Cooperation Programme.

The 4,000 or so advisers, consultants, and associates FAO had annually on field assignments in the early 1980s may be set against a similar total of professionals and general service staff at its headquarters. Recruiting, paying, and supporting technically the field staff requires a lot of headquarter staff. At the same time they assemble information, analyse experience, and produce direct advice and assistance through publications, field visits, and the organization of technical and policy exchange.

After some experimentation, FAO became highly professional in its management of field projects. This is evidenced by the large number of bilateral aid agencies that look to FAO to run projects they finance. The main unit concerned is the Agricultural Operations Division in the Agriculture Department. Similar units in Fisheries and Forestry manage their own field projects. The operations staff brief field advisers administratively and arrange for their introduction to the government concerned. They refer them to the appropriate professional units for technical briefings and channel back to them subsequently their comments and advice. This division of responsibility has proved itself well – subject to occasional delays due to the absences of strategic individuals on leave, sickness, and professional travel.

ALLIES AND OVERLAPS

The development assistance field is crowded. This is especially true of the main areas of interest to FAO, food and agriculture, less perhaps of fisheries and forestry. FAO draws the bulk of its resources for action projects at the country level from UNDP and some individual government aid programmes. Its partnership with the World Bank and IFAD puts capital support behind its technical recommendations. The CGIAR international research chain should provide new material for its production improvement recommendations. There are other international and national aid agencies with which FAO sees areas of duplication, agencies that it tries to control, and still others that try to control FAO.

United Nations

The General Assembly of the United Nations in New York is the ultimate authority in the UN system, and in general FAO is under obligation to follow its recommendations. Through its Economic and Social Council the UN exercises a broad co-ordinating role over all the UN agencies. DG Sen opposed this on the grounds that it would stifle their initiative and freedom of action. In 1977 a UN Department for Development and International Co-ordination with a head salaried at over $100,000 per year was established. His role was to exercise a continuing co-ordination of the UN system. This was still more resented at the top of FAO.

The United Nations Development Programme, long headed by Bradford Morse, is located in another building on the UN Plaza, New York. Its primary role is to mobilize aid funds from UN member countries and allocate them to individual developing countries according to their criteria of need, and to the various UN specialized agencies by their subject specializations. At the first UN technical assistance conference in 1950 fifty governments pledged $20 million. FAO received the largest share (20 per cent) because of the prime importance of food and agriculture. UNDP continued to be FAO's main source of funds for technical assistance to individual countries or to developing regions.

The United Nations has Regional Economic Commissions based in Addis Ababa, Baghdad, Bangkok, Geneva, and Santiago, focusing on overall economic development. Each has a substantial staff grouped into subject matter divisions, including one for agriculture. To avoid duplication and to take advantage of the opportunity to address ministries of economic development and planning, FAO contributes a Chief and one or two professionals to these agriculture divisions. They channel information, study materials, and experience into the Economic Commission which is already available in FAO. Agreement for this with the UN Economic Commission for Asia and the Pacific (Bangkok) was interrupted following an encounter between DG Saouma and his counterpart executive secretary. This UN Commission subsequently initiated fertilizer and agricultural marketing activities fully overlapping that of FAO.

Together with other UN agencies, FAO contributes to the financing of a Joint Inspection Unit. It examines areas of FAO work in turn; information has to be provided to it on request, and criticism in its reports is transmitted back to the units concerned. In general it is not considered to be very effective.

FAO follows the UN scale of salaries by grade. Its staff participate in the UN Pension Fund.

World Bank

Located in Washington DC this is the main international source of development capital. Initially it did not lend much to agriculture. Following its 1964 agreement with FAO over the preparation of projects for its financing, loans to agriculture increased rapidly. They were further expanded through the International Development Association (IDA). This channelled World Bank earnings through interest on loans, plus additional pledgings by member countries, to countries unable to bear the cost of conventional loans. Total Bank investment in agriculture went up from $150 million annually in the 1960s to $4 billion in the early 1980s.

The share of projects prepared by the FAO/World Bank Cooperative Programme in total World Bank lendings has gone down in recent years with the expansion of 'structural adjustment' lending. The Cooperative Programme, however, still functions. Similar arrangements hold good with the African (Abidjan) and Asian (Manila) regional development banks, and with IFAD.

International Fund for Agricultural Development

IFAD is an independent specialized agency of the United Nations. It was set up in 1977, an outcome of the 1974 World Food Conference, with the objective of mobilizing additional finance for agricultural and rural development. It was a device for trapping part of the riches flowing to the petroleum-exporting countries from the sharp price rises of 1973 and after. If as a group they would provide 50 per cent of its funds, the industrial countries would match it. Peter Gill saw their contributions as a penance for the damage they inflicted on fragile Third World economies (Gill 1986). Conventionally, the president of IFAD comes from one of the OPEC countries.

IFAD differs from the World Bank in that it makes grants as well as loans; it focuses specifically on small-scale farmers and does not apply the 12 per cent internal rate of return criterion commonly used by the World Bank. This suited the mid-1970s mood of helping the poorest of the poor. It has only a small permanent staff, 140 in total. Project supervision is entrusted to the World Bank or to one of the regional banks who are often co-financers.

IFAD uses rented accommodation on Via del Serafico, off the Via Laurentina, Rome. A shuttle bus is available from FAO. Originally it was to look to FAO for project ideas. Its 1988 report notes with satisfaction, however, that twenty-two out of twenty-three projects approved that year were IFAD-initiated. IFAD's leadership attaches a lot of weight to its 'specificity', i.e. increasing food production, alleviating poverty and securing the active participation of the people concerned.

By the mid-1980s IFAD had put $2.4 billion into 220 projects in eighty-nine developing countries. They were instrumental in raising cereal output by 24 million tons. It has shown that if the poor are consulted about it first, they can use additional funds to great advantage. IFAD has good public relations; even so, the funds available to it in 1989 were in real terms only half what they had been ten years earlier.

World Food Programme

This is another organization located in Rome, allied to FAO but restive for independence. Its purpose is to channel to good use in the developing countries food that is surplus to commercial requirements in other parts of the world. It began with an offer by the United States to pass through it, under its law PL 480, food valued at $40 million. By the early 1980s the resources pledged to it annually in the form of surplus commodities, shipping, and cash, were approaching $1,000 million. Of the food made available to the WFP, 20 per cent can be allocated by the FAO Director General at short notice to meet an emergency. The balance has to be assigned under projects approved by its governing body. A main concern of the member governments is to ensure that these projects actually contribute to development and are not simply a device for replacing commercial purchases. Countries such as Australia and Canada, for which grain exports are a main source of revenue, are particularly exercised about it.

Mordecai Ezekiel of FAO wrote an influential paper showing how surplus food could contribute to development. It could be used as wages to people employed in building and maintaining rural roads, irrigation and drainage ditches, and those employed in planting trees and shrubs to fix sand dunes. It could be an inducement to children and adults to undertake training to improve their knowledge and skills. The range of uses acceptable to WFP has since been extended considerably to include, for two countries in 1987, relief during a foreign exchange shortage.

WFP is run by an executive director, in 1990 James Ingram of Australia. His governing body is the Committee on Food Aid Policies and Programmes. Its members are elected half by the FAO Council, half by the Economic and Social Council of the UN. WFP, with a staff of

300, operates from rented offices in the Via Cristofero Colombo, Rome. When first established it provided job opportunities for many people who felt they were not appreciated in FAO. It is now stronger and has its own administration.

The World Food Programme continued to grow through the 1980s when many other aid agencies faced donor reluctance. It had the advantage of serving several complementary interests:

1 Through it, governments can dispose of burdensome surpluses generated by their own political pricing.
2 Some governments contribute shipping; this provides work for their ships and merchant marine.
3 The food provided to meet recognized shortages goes to the governments concerned; they can sell it and apply the proceeds in local currency to uses which they determine.

Food aid, for which WFP is a major vehicle, has known disadvantages:

1 Its release on developing country markets can depress prices to local producers. Illustrative is a case where a domestic dairy group in Bengal, helped under a voluntary aid programme, was forced out of business by co-operatives retailing milk made with powder imported as food aid.
2 Its availability eases the pressure on recipient governments to promote their own agriculture by incentive pricing, improving the supply of fertilizer, etc.
3 Food aid can promote consumer tastes in Third World countries for items that can never be produced there economically – typically wheat flour in the tropics – and so increases dependence on imports.

Concerned primarily to move food surpluses out from expensive storage, WFP's major donors have not watched too closely how they were used. In some countries, diversion and sale for personal gain has been endemic. An evaluation officer told to keep quiet about this while he was with WFP, damned it in the London *Observer* a few days after he left.

World Food Council

A handicap for FAO in mobilizing the full weight of the developing countries behind the conquest of hunger has been that it addressed only their ministers of agriculture. In most cases FAO was already preaching to the converted; but the ministers were unable to carry their respective governments with them. They faced urban bias in the educated minority, leaders fearful of riots in the capital, and political ideology blinkered to the realities of agriculture production and marketing. Some years ago the Minister of Agriculture of Madagascar was seeking aid from external donors working together as the 'Paris Club'. For it to be effective, they said, his government must lift off the backs of its farmers parastatal monopolies stifling incentive to produce more food. The Minister agreed.'When will it be implemented?' he was asked. 'As soon as possible,' he said,'if I am still Minister when I get back.'

The World Food Council, set up in 1974, was intended to secure a higher political focus on food supply and distribution problems. It was sponsored by the United Nations as a whole. It would not duplicate FAO, but rather reinforce it. For information and analysis it would look to FAO's economic and technical divisions. Staffed only with a small secretariat to service its meetings, it is located on one floor of FAO Building A. The executive secretary in the late 1980s was G.L. Trent of the Canadian Department of Agriculture.

Its assessment for the sixth ministerial session at Arusha in 1980 of food distribution programmes for lower income consumers was timely and well done. It could have been done equally well, however, by FAO's Nutrition and Food Policy Division. While the mandate is for close collaboration, FAO senior management is inclined to resent WFC as 'second-guessing' its conclusions and recommendations.

These are only a few UN agencies involved in food, agricultural, and related development issues. To meet the evident need for co-ordination, a host of inter-agency committees have been established. There are also agreements defining areas of responsibility between agencies. FAO has signed twelve of these. The overlapping which led to them still continues, however. It could be ended by informal talks between agency heads, but alas they do not speak to each other.

International research institutes

To bring the resources of modern biological and socio-economic research to bear on subtropical and tropical agriculture, eleven international institutes were established in the 1960s and 1970s. The International Rice Research Institute (IRRI) in the Philippines and the International Maize and Wheat Improvement Centre (CIMMYT) in Mexico developed the higher yielding varieties of rice and wheat that sparked the 'green revolution'. The International Institute of Tropical Agriculture in Ibadan, working on food crops in Africa, has still to

31

achieve such a breakthrough. An International Food Policy Institute was set up in Washington following the food crisis of 1973. In 1989 it had a technical staff of eighty-five including research fellows and research assistants. These institutes all come under the Consultative Group on International Agricultural Research, made up of FAO, the World Bank, UNDP, and thirty-two governments, international and regional organizations, and private foundations that contribute to their funding. FAO provides the secretary of CGIAR's Technical Advisory Committee.

Bilateral and non-governmental agencies

FAO assistance proceeds alongside official and other aid which dwarfs it in scale and sometimes in influence. There have been countries where American or French advisers had the ear of the government; various African governments, from Ghana under Nkrumah, to Ethiopia have focused on large state farms under 'eastern bloc' tutelage. Nordic aid maintained fifty co-operative advisers in East Africa for over a decade, far exceeding the intensity of FAO assistance. There are also religious and other voluntary groups trying to go directly to the people.

The impact can be negative; the diversity of training provided to co-operative personnel in Zimbabwe became a source of confusion. Installation of over twenty different makes of irrigation pumps in Kenya complicated maintenance. One overloaded ministry in Burkina Faso had to accept over 300 requests for appointments from aid representatives in a recent year. There have been many attempts at co-ordination – the dispatching of aid programming missions, meetings of centrally placed representatives as in the 'Paris Club', informal arrangements at the country level. Collaboration can be achieved on specific defined projects (the building of roads in Nepal, for example). More generally, all parties – the recipient government and the aid agencies – want to keep their options open.

3

DIRECTORS GENERAL

The first Director General of FAO was Sir John Boyd Orr, knighted in 1935 for his services to UK agriculture. By profession a nutritionist, he had been head of the Rowett Research Institute near Aberdeen in Scotland. He was also Professor of Agriculture of Aberdeen University. He resigned both posts in 1945 to stand for Parliament and was duly elected. Technical adviser to the British delegation at the founding conference of FAO in Quebec, October 1945, he was elected there to the post of Director General of FAO.

Boyd Orr was a man of great energy, drive, and enthusiasm. His objective for FAO was to make it a world food board which would ensure a better distribution of food in the world. It was said of an intervention he made during the third session of the FAO Conference in Geneva, 'The Director General rose, with his face white with anger. He had the fire of God in his belly and he belched.' He established the FAO emblem – the letters FAO around a head of wheat. It was first used by a Danish silversmith on a badge for participants at the Second FAO Conference at Copenhagen in September 1946. Boyd Orr added the motto Fiat Panis (Let there be bread).

With his interest in food distribution to the fore, Boyd Orr set up a strong Economics Division to handle agricultural economics, statistics, and commodity issues, and assigned to it the bulk of his resources. There was difficulty in getting him to look at the technical side of agriculture. Establishment of a division for this came under pressure from a man in the UN Relief and Rehabilitation Administration who wanted to be transferred to FAO, together with some UNRRA funds. When finally asked to approve a plan for an Agriculture Division, Boyd Orr drew a red pencil through the proposed section on extension, with posts for four professionals, saying 'We can hire one man and a secretary for three months to write all that FAO will ever need on agricultural extension' (Phillips 1986).

Sir John Boyd Orr had a casual approach to budgets and work programmes, to the point that his Committee on Financial Control became restive. A lot of time went into preparing a detailed staffing plan and budget. When he looked at the eventual draft in terms of what could actually be done, he wrote a new set of figures on the back of an envelope and pushed it over to the Chief of Budget and Finance saying, 'There's your budget.'

To me, Sir John Boyd Orr was the familiar photograph on the wall with a pipe in his mouth. Others remember his habit of tugging at a shaggy eyebrow when pondering a problem. He maintained the English tradition of having tea each afternoon with members of his staff. He left FAO a deeply disappointed man. His plans for a World Food Board had been set aside by his member governments. He also wanted to provide developing countries with pumps, tractors, etc., via loans from the World Bank. This was rejected by his Council on the grounds that they were not engineers and so not competent to deal with such matters. So his positive proposals came to nothing; all that was left of them was the Committee on Commodity Problems and an obligation to 'keep under review the state of food and agriculture in the world'. These remained major activities of the Commodity and the Economic Analysis Divisions respectively throughout the 1960s. Lord Boyd Orr, as he became in January 1949, received the Nobel Peace Prize that year.

Boyd Orr left FAO after his term in office, excited that he would then be going into business. The next DG, Norris Dodd, was quite a different man. Trained as a pharmacist he had already run a range of businesses – pharmacies, a farm, a sawmill, feed supply, milk distribution, and a forage warehouse. He moved into agricultural administration through county and regional agricultural adjustment committees in the 1930s to become a director in the Production and Marketing Administration of the United States Department of Agriculture. He was Under-Secretary of Agriculture when elected Director General of FAO in 1948.

He had first-hand experience of how to run a business, and a government agency, and was considered a hard-headed administrator. He irritated Sir John Boyd Orr when presented to the senior staff as incumbent DG, by pursuing a direct enquiry into their work. This provoked a sharp 'I'll have ye know that I am still the Director General' from Boyd Orr. Ralph Phillips saw Norris Dodd as a quick-witted and effective director general who ran a tight ship. He recounts an occasion when the Frenchman in charge of protocol complained that the travel programme set up for him along with the DG was much too rigorous.

Mr Dodd's comment was, 'What is the matter? Don't you think you can take it? If you can't you can stay at home.' So Veillet-Lavallée went along. To a staff member of contemplative mien, Dodd remarked, 'If you want to get on in this organization you've got to keep jumping.' The man asked him 'How do I know which way to jump?' to which Dodd replied, 'Does not matter; just keep jumping.' Flying over the Brazilian jungle to a meeting in Buenos Aires he remarked to his companion, 'Got a brother down there; he couldn't draw fast enough.' In his office in Rome he kept a fly-swatter on his desk and would interrupt a discussion to swipe at one that dared to come within reach. While vigilant enough over FAO's expenditures, he was quick to recognize an immediate need. In the face of a locust outbreak in the Near East in the early 1950s there was only $40,000 in the budget. When asked for substantive help he put his feet up on the desk and said, 'Then make it $200,000.' In this way, FAO began one of its most effective practical programmes.

About the time that I was considering joining FAO I saw a picture of Director General Dodd holding a plough. I commented to an American professor friend that it must be good for FAO that its director general had himself been a farmer. I am still inclined to that view. His response, however, was that 'if Norris Dodd has his hands on a plough then it must be that there are some photographers around.'

Dodd was always a practical man. In Rome, his secretary was installed in an apartment just above his own. As his wife it would have been difficult to keep her on the payroll; they married the day before he left FAO and she qualified for a survivor's pension.

The third director general, P.V. Cardon, came from Utah. While his career was primarily in research, his degrees were in agricultural economics. For three years from 1922 he edited *The Utah Farmer* in Salt Lake City. He rose through administration of agricultural research to become head of the Agricultural Research Administration and then of the US Department of Agriculture Graduate School.

Cardon was a man of high intellect and ideals, too sensitive, however, to withstand the pressures of managing a large public organization. He had already been ill for this reason in his previous post. Ralph Phillips remembers his concern on hearing Cardon say during his acceptance speech to the FAO Conference, 'When decisions are needed I shall make them'. This seemed to indicate difficulty over what should be automatic in a man taking up such a post. Cardon resigned in 1956, following breakdowns attributed to constant needling over budgetary matters by representatives of his own government.

B.R. Sen, the fourth director general, was educated at Calcutta University and Oxford. He had risen through the Indian Civil Service from District Magistrate to Director General of Food and to Secretary, Ministry of Agriculture. He had also been Ambassador to Italy together with Yugoslavia and to Japan and the United States. He was put up as candidate at the suggestion of the British Government; it saw a need for change after two Americans.

Coming from one of the largest developing countries, with direct experience of the Bengal famine, helped Sen to seek financial resources for FAO that DGs from already developed countries would not have envisaged. He raised FAO's two year budget from $13.5 million to $167 million. His tenure of office coincided with the coming to independence of many African territories formerly approached only through a metropolitan country. It saw large-scale funding of technical assistance through FAO by the United Nations Special Fund for Development. Sen called the first World Food Congress in 1963. He initiated the preparation of an Indicative World Food Plan and the FAO preparation of investment projects for the World Bank. The FAO Freedom from Hunger Campaign, bringing to FAO voluntary funding from non-government organizations and the collaboration of big fertilizer, food, and agricultural machinery industries was also launched by Sen. He saw this mobilization of voluntary assistance as his greatest achievement.

Once appointed, Sen left no doubt that he was in charge. Dr Wahlen, as chairman of a staff morale committee, let him know that the committee was available if he needed advice. Sen responded, 'So long as we have a morale committee we will have a morale problem. The committee is abolished.' He also dismissed forthwith a staff spokesman who used violent language against the FAO administration.

Self-consciousness and shyness, he says, led to his adopting a mask of detachment. To others he seemed distant and unapproachable. A colleague called to his office came back saying that he 'had been talking to God'. Sen could, however, express appreciation. A supportive statement to a committee on the Freedom from Hunger Campaign brought him down from the podium to walk the length of the room and shake hands with the speaker.

Sen did very well for FAO. He regained the initiative over FAO's operating budget that had been lost by his predecessor. While he enjoyed exercising authority, nobody said that he abused it. He delegated responsibility. He kept himself informed through daily 'morning meetings' with his ADGs, and one attended by directors or their representatives every two weeks. He instituted a 'monthly letter' to ministers of agriculture on FAO's plans and activities. He faced up to the Vatican on the birth control issue. Against its view that there was the resource potential to support larger populations he stressed that these resources were not available at particular times and places. B.R. Sen had the background, vision, and leadership capacity, to raise FAO to its greatest height.

Frank Weisl, long-time ADG for Administration, who had served under three previous DGs, once told me that 'Sen made only one wrong decision – to put his country representatives under the UN resident representative.' Effectively, he was deprived of an independent voice with the government. There was, however, a significant saving in cost; and with the establishment of its Special Fund for Development, UNDP had become the vehicle for a tremendous expansion of FAO's field operations. DG Saouma made a point of reversing this decision. Some of the sharpest criticism he has received has been over the cost of maintaining a large number of independent FAO country representatives and the use he made of them.

A.H. Boerma was Director General of FAO from 1967 through to 1975. He graduated from the Agricultural University, Wageningen where he specialized in horticulture and agricultural economics. He worked first with a Dutch farmers' organization then as the government officer in charge of preparations for food distribution in the event of war. He ended up as Acting Director General of Food. In 1946 he became the Netherlands Government's Commissioner for Foreign Agricultural Relations. In 1948 he says he exchanged posts with another prominent Dutchman, S.L. Louwes, and became FAO's Regional Representative for Europe, based in Rome.

It was because of an address Boerma gave to the University of California in 1954 that I joined FAO. My next contact was on an overnight flight to Bangkok. We were going to a meeting on rice supplies and pricing. He wanted somebody with good English to keep notes and draft a report. I sat next to him on the plane, studying the technical background; he read a paperback titled 'Joy Street'. At Bangkok we went directly from the airport to the meeting. He delivered a resounding opening speech without reference to notes; then he said he would go to the hotel to get some rest, leaving me to carry on the meeting.

In 1958 he became Director, Programme and Budgetary Service, a post raised to Assistant Director General rank two years later. Like Boyd Orr he saw a role for a world food board and became first Executive Director of the World Food Programme when it was established in 1962. Whereas Boyd Orr was never seen without a pipe, Boerma favoured a big cigar. It amplified his personality. He was a superb public speaker in a range of languages. Each statement came forth with a smile counting, as it were, on audience endorsement forthwith. An easily approachable consensus man, he was inclined, it was said, to agree with the last person who had his ear. This did not make him a good Director General. The World Food Congress held at The Hague in 1968 coincided with the peak of 'student power'. It was lost in verbal wrangling with young people's representatives. The Conference following the 1973 crisis in world food supplies was held in Rome, not by FAO but under the auspices of the United Nations, New York. New organizations were set up to take on responsibilities that logically would have gone to FAO, if it had been in favour. The crisis opened the way for FAO to put forward a new ambitious programme. This was left to the existing internal units; they came up with a lot more of the same thing. With remarkable prescience Boerma's assistant in Programme and Budget, Jean Fairley, wrote in 1961:

There once was a Dutchman called Boerma,
Who spoke in a small, polite moerma,
In view of his size
This caused widespread surprise:
One expected a loud voice, and foerma.

Edouard Saouma received a degree in agricultural chemistry from St Joseph's University, Beirut, and a diploma in agronomy from the National School of Agronomy, Montpellier. After heading an agricultural school and then a farm mechanization centre in Lebanon, he became Director General of its National Institute of Agricultural Research at the age of 29. He attended FAO Conferences as a delegate of the Lebanese Government. In 1961 he was largely instrumental in swinging six Arab votes in favour of Sen. Shortly thereafter he was appointed Deputy Regional Representative of FAO for Asia and the Pacific based in Delhi. He became Director, Land and Water Development Division in Rome in 1965 and Director General in 1976. Saouma came in as DG with a number of points in his favour:

1   He had worked in the organization recently as director of an active technical division. He knew exactly how it operated, who did useful work, who did not, and what support it received from FAO Administration.

2   Though a Maronite Christian, as an Arab from Lebanon he was elected with the full support of the Arab world. His election took

place at the peak of the petroleum boom. If led by Saouma it was expected that FAO would receive generous financial support from the petrol-rich Arab governments.

His initial steps as DG were wise, timely, and well-received. He announced to the staff that he did not intend to compete in a popularity contest. This was salutary and appropriate. A number of FAO people had coasted along, doing the minimum for the organization, making the most out of it. They thought they were protected by the continuity of their employment and the staff union. They were now faced by a DG who was prepared to fire them on the spot if he thought he had grounds for it. Some of those so treated took their case to the ILO tribunal in Geneva and had to be reinstated or paid compensation. But he could still put somebody else in their job. They would be left with a salary, but without anything to do until they gave up and took another post. Some of these events had a comic twist. Saouma made up a quarrel with the Argentine head of Plants Division so that he had to resign. As division directors there had been a game between them over answering the telephone. Each instructed his secretary to get the other on the line. The effect was that for some time neither had been able to speak to the other. The disappearance of this man may not have been much of a loss to the organization. He was replaced by another Latin American, previously Minister of Agriculture for Mexico, of German and Amerindian descent, practical and without pretension.

There was no question of strike action under early Saouma. A man chained himself to a machine in the printing shop in protest against a well-earned dismissal. Photographs appeared in the newspapers. Apparently FAO's Italian guards had been reluctant to keep out the press and remove the offending staff member. Shortly afterwards the FAO corps of guards took on an Asian hue.

He threw out the Programme of Work presented to the 1976 FAO Conference by his predecessor. This document took advantage of the international climate created by the food shortages of 1973–74 and the subsequent World Food Conference to secure a large increase in FAO's budget. It proposed, however, only an expansion of the existing programme with no new ideas. Saouma secured a revision that cut down sharply the increases in headquarters staff, meetings, and publications, diverting some $21 million into funding direct assistance to member countries through a Technical Co-operation Programme. He also initiated the establishment of a corps of FAO country representatives, fifteen initially, to be increased to sixty-two. It was presented as a decentralization of FAO's resources in the interests of closer contact with national governments. He foresaw an expansion of FAO's resources for technical assistance from bilateral sources, and needed representatives independent of the UNDP to manage their use. Presentation of this new programme was a coup. It earned him the following comment in the London *Economist* under the heading 'FAO fight against flab':

> By cutting back sharply on meetings and publications, Mr Saouma has freed $18.5m from the agency's two-year budget of $167m: this he intends to use for a new technical co-operation programme. Since his appointment last November he has reviewed proposals for the creation of 519 new posts, and abolished 330 of them. Meetings that were expensive in flight, hotel and interpreter costs have been cut back from 408 to 253 for the current two-year period.
>
> A similar review of publications has cut the total from 298 to 204 for a saving of $1.3m.
>
> Mr Saouma reported to the council that the FAO's burgeoning staff 'tended to propose an ever-growing number of missions, to create different kinds of masterplans and studies, to attend more meetings to discuss them and to produce more documents to describe them.' Training tended to concentrate on high-level seminars in capital cities rather than at farmer, fisherman and forester level. He was anxious to ensure that meetings and publications avoided 'academic, diffuse and self-perpetuating activities in favour of ones promising concrete results of immediate value to member nations.'
>
> The decision to place more emphasis on field work is undoubtedly a wise one.
>
> Comparison with the American department of agriculture's staff of 100,000 makes FAO seem quite a slim creature. Natural wastage and the new firm hand of Mr Saouma may get it back into even better shape in the next few years.
>
> (7 August 1976)

Criticisms of Saouma in action have been:

1 The combination of the Technical Co-operation Programme and FAO country representatives could be highly positive. The representative in Tanzania in the mid-1980s earned full admiration from the World Bank, for example, for his ability to draw on TCP funds promptly to meet an immediate need for assistance, then

mobilize bilateral funds channelled through FAO to follow it up. However, it was often said to be used by Saouma to gratify ministers and secure support for his re-election. Approval of TCP projects was kept in his hands; it could go to proposals seen by his own technical officers as a waste of time and money. The scope for mobilizing bilateral support in countries such as Tanzania and Peru was substantial, but what could an FAO representative do in Uruguay beyond reminding ministers periodically that Mr Saouma was their friend? While Saouma did not prejudice performance at headquarters by inappropriate staff appointments, his corps of country representatives might afford ample scope for absorbing relatives and persons who had acted favourably or whose governments promised to do so.

2 The process of fulfilling a single-minded ambition often enables a man to control, or at least to conceal, certain characteristics. It is only when the ambitions are fulfilled that these characteristics can become evident. This has been the case with Saouma. As Director General it has been said that his arbitrary manner prejudiced staff morale and FAO's ability to employ the best people in its technical fields. This was a major issue; for those whose position put them in direct contact with Saouma, the expression of a contrary view could be interpreted as a personal affront or disloyalty.

3 Saouma's acute sense of his own personal importance is seen to act negatively for FAO, notably in the northern countries. His refusal to visit Japan unless he was to be received by the Emperor Hirohito was one example. 'The Emperor only receives heads of state', was the Japanese reply; but Saouma won.

4 The governing body of FAO felt that his predecessor travelled too much in pursuit of his re-election at the end of a four-year term. It changed the constitution to provide for one term of six years with no re-election. Saouma had this reversed to permit re-election for terms of six years. He is now on his third term, making a total of eighteen years. In the FAO lift one day a colleague pointed to a babe in arms. 'One day he will be a valued staff member of FAO,' he said. 'He may well see another director general' was the spontaneous come-back. 'Magari – if only' was the muttered accolade.

THE ELECTION

Directors General are appointed by the FAO Conference of government delegations. Candidates must be nominated by their government. If there is more than one there is an election. For election a candidate needs more than 50 per cent of the votes. After his first term, Norris Dodd did not stand again because he did not expect to be nominated by his government. He had been a protagonist of the Democrats, who had been succeeded by the Republicans with Eisenhower as president. So a new American candidate, Cardon, was presented.

The election of 1956 following Cardon's resignation was one of the most exciting. There was an American candidate, J.H. Davis, who after a second ballot still lacked one vote. Since he was evidently not attracting overwhelming support he decided to withdraw. This left the third ballot to Sen and Mansholt of the Netherlands. Sen won by 42 to 29.

Until Sen was elected it had almost been assumed that the DG would come from one of the countries making a major contribution to the FAO budget; that this and other such posts in the UN agencies would be shared equitably between them. It now became clear that to be elected, a candidate must attract the votes of the developing countries for, from the 1960s, they became an increasingly overwhelming majority.

At the next election, Boerma stood against Santa Cruz of Chile, who had been the FAO regional representative for Latin America. This resolved into a line-up of Latin American and francophone countries for Santa Cruz, north European and anglophone Africa and Asia for Boerma. The critical issue was which could attract the Arab vote, now around twelve countries. Boerma succeeded, reputedly by promising to eliminate the strong internal Jewish influence which had developed under Sen, termed the 'Senhedrin' after the Jewish council of high priests.

At the 1975 election Saouma stood specifically as candidate of the Arab countries with Saudi Arabia furnishing financial support for his campaign. He was lucky over this: the conflict between Moslems and Christians in Lebanon was just coming to a head and he was a Lebanese Christian.

As a French speaker, educated in France, he also commanded the French African vote. Pathetically, one of the two anglophone candidates took lessons in French to better present his case. In any event they split the opposition to Saouma who won easily. The consequence was the appointment of Frenchmen to a number of key positions and a sense amongst staff from other backgrounds that fluency in French was essential for the inside track.

When next the anglophones fielded a candidate in 1987, they went the whole hog and proposed Moise Mensah. He was from Benin, a francophone and left-oriented African country. He had been FAO representative for Africa and was currently Vice President for Operations in IFAD. The Canadians circulated a video film detailing Saouma's style and practice. There were accusations that Saouma promised lucrative projects under his Technical Co-operation Programme and posts for those who helped to secure votes. Lobbying representatives personally, he won over not only enough of the small countries but also France, Germany, Italy, and the east Europeans. Cries that his technique was corrupt went unheeded. The next election will be in 1993. Candidates must be nominated by April of that year.

DEPUTY DIRECTORS GENERAL

The scope left by the Director General to his deputy has varied greatly with the personalities concerned. 'Sir Herbert Broadley maintained continuity through several DGs in the 1950s. He had been knighted for his services to the Food Ministry of the UK during World War II. In FAO he was noted for his masterly expositions of FAO policy and quick assessment of a complex issue. Broadley worked so hard keeping FAO going that he never had a chance to visit the countries he was working for. He lived in the Hotel de Ville for most of his stay in Rome. Broadley could take some original initiatives. One Saturday morning I arrived at FAO to work quietly on a book. I found that it had been rented for the day to a film company. They liked the spacious entrance and had Building A signed 'Grand Hotel'. International stars rode slowly up the drive; they gaze longingly at each other in the doorway.

O.V. Wells had been head of the Agricultural Marketing Service of the US Department of Agriculture. He had the same gifts as Broadley. Under Boerma he had an important influence on FAO's technical programme. On a warm afternoon he could sleep through a lengthy speech by a Conference delegate, wake up when the voice stopped, and make prompt recognition of the speaker's contribution. Wells talked of a hard upbringing – the local sport for youths in his neighborhood was rounding up young cattle and putting on a brand before others could claim them. He worked on the railroad to pay his way through college, and when he came to Rome he had already earned a government pension. He played golf, ate at good restaurants, and put a lot into FAO.

R. Jackson, DDG 1972–7, previously head of the Fisheries Department, was largely left aside. One of Saouma's innovations was to establish his own 'Cabinet' with a chief and several assistants. They handled much that would otherwise have gone to a Deputy DG. It was the same with R. Phillips (1978–81). Apart from representing the DG when he was away, Phillips undertook only a few specifically designated assignments.

E. West (1982–6) made a hit as UK delegate to the governing Conference. Each year he brought a new joke. He was useful to Saouma because he could intimidate delegates to FAO committees and conferences with a blend of fore-knowledge and arrogance. He had imbibed the British Civil Service tradition of proving black is white if the minister says so. Those who accused Saouma of corruption condemned West also for his condonation. They also saw the 'Senhedrin' risen again.

All of these men had a depth of background in FAO and in their own government services. West's successor, D.J. Walton, was a professional speech writer, first for Boerma, then for Saouma. His appointment to deputy director general brought the post to its minimal status.

Although he was Deputy DG for less than a year, the most eminent man FAO ever had at this level was undoubtedly F.T. Wahlen. FAO has had on its staff a number of men who had been ministers in their governments, less often men who went on to become one. Dr Wahlen was the only one who left FAO to be a member of his country's governing council and then its president. Phillips, who worked under him in FAO, tells of an occasion when Wahlen, in office, visited him at the UN buildings in Geneva. The door flew open and a guard announced 'His Excellency, the President of Switzerland'. Wahlen came in and said 'Shall we go for a coffee?'

Wahlen had learnt Russian as a student and spoke German, French, Italian, and English. In Canada he ran seed laboratories for the Dominion Government. He went back to Switzerland as director of an agricultural experiment station then became Professor of Agronomy at the Zurich Polytechnicum. It was from this post that he was drawn into the management of food supplies during World War II, implementing as Commissioner for Food Production the Wahlen Plan for which he became a national hero. Domestic production of grain and potatoes was pushed to the limit.

Wahlen came to FAO as the director of its Agriculture Division in 1949. It was with Wahlen as its technical leader that the organization's professional flag flew highest. FAO was still small at that time. Wahlen was world-renowned. He could supervise his staff personally. He was known for his 'liquid lunches', glasses of grappa, the drink of the Swiss farmer, held once a month for all-comers. He was a serious man, but he liked people to relax and enjoy themselves.

Wahlen had a healthy disrespect for economists. At one of his informal meetings he commented to someone who said that to carry out a certain programme they would need a hundred agronomists, 'For a moment I thought you said economists.' Ironically, it was economics that later brought him unpopularity with farmers in his own country. The burden of subsidies had to be limited. Cartoons on carnival floats had him cruising over the Valais farming area in a helicopter, spraying unauthorized vineyards with defoliants, and levelling a blunderbuss at potato-growers who had planted beyond their quota. All of these drawings featured a domed professorial head, a determined look, and deep, prying eyes.

Earlier, Wahlen would have liked to be director general of FAO. He did not seek nomination when Cardon resigned because he foresaw another American candidate and did not want to stand against him.

A CONTROVERSIAL ASSISTANT

For a period in the mid-1960s FAO had its *éminence grise* in Egon Glesinger. He had a hold on Director General Sen. He was able to talk easily to a man who to other senior officers seemed distant and difficult to approach. Through a contact with the Swedish development economist Gunnar Myrdal, he dangled before Sen the prospect of nomination for a Nobel prize.

Glesinger had risen in the FAO Forestry Division to become its director. He obtained authority from Sen to call on the staff of other technical divisions to give priority to work on his Mediterranean Development Project. Eventually a new post was established specifically for him – Assistant Director General for General Affairs with special responsibility for the Mediterranean project. The concept was attractive. Selected poor areas in countries around the Mediterranean would be helped to plan an integrated development. They would become models for replication. Funds for this were available from the UN Special Fund for Development which had just been established. Hitherto, the various technical divisions had tended to propose to it projects featuring their own professional interests. These area development projects called for a combination of disciplines in effective interaction. Plant production and protection, animal production, irrigation, provision of credit, fertilizers and other inputs, organization of marketing to assure a financial incentive to farmers – all these were needed within an overall investment frame. Glesinger wrote a technical justification for his Mediterranean projects in *Scientific American*. He mobilized support from the governments concerned and from the UN Special Fund. He did very well in bringing together professional expertise from different parts of FAO. Later, this became standard practice, with a new department established to ensure its implementation. Over-institutionalized in some of the integrated rural development projects of the 1970s, it eventually reached a wasteful extreme. Glesinger became highly influential, but never liked. There was a lack of rapport with colleagues and of generosity over small things. It gave many of those who worked with him the feeling that they were being used, instead of being engaged in the joint pursuit of a development goal.

It was a precarious regime. The developing countries of the Mediterranean are predominantly Arab Moslem. The managers of the development projects in their interest were Jews. The first action of the next director general was to suspend the Mediterranean project and put aside its leader.

4

PROFESSIONALS

The professionals in FAO are staff members with posts graded P1 to P5 in accordance with the United Nations system. We may also class with them director grades D1 and D2, to which many professionals aspire. The basic qualification of a professional in FAO is a university degree, or its equivalent in formal training. A candidate should then have relevant experience on a scale rising with professional level – very little at P1, rising to 5 or 7 years, and eventually 10 years at P5. Further qualifications required are the ability to speak and write in at least one of the FAO's official languages, in particular English, French, or Spanish, and for the higher grades evidence of an ability to work independently, supervise others, and assume leadership responsibilities according to the level of the post. In relation to university appointments, a P5 in FAO would be comparable to a professorship, P2 and P4 to assistant and senior lecturer, respectively.

POSTS, GRADES, SALARIES

Salaries for professionals in FAO, according to the scale in effect from 1985, began at $17,000 for P1 rising with ten annual steps to $22,000. The P5 salary was $36,000 rising to $43,000. A D2 started at $45,000 and ended with $48,000. Assistant directors general received $54,000 with no annual increases, the director general $78,000 with allowances of $57,000.

These figures are for individuals without dependents. There is a family allowance which adds $1,000–$3,000 annually, according to the level. By agreement with member governments, all these salaries are free of income tax; or if this is levied, as by the government of the USA, it is then reimbursed by FAO. There is provision for salary adjustment according to the cost of living. Rome is not considered expensive by

Geneva standards, but FAO personnel working in the Gulf countries, for example, would receive substantially more. There are also allowances to help cover the cost of educating the children away from home, to cover home leave travel for the family every second year, etc. Staff may participate in a health insurance programme with pension entitlements, making regular contributions that are matched by the organization.

While FAO salaries generally permit a comfortable standard of living, their level can arouse strong feelings. Many of the people who vote for aid to the Third World on grounds of conscience and principle see FAO salaries as excessive. They contrast sharply with earnings in the countries to be helped, with religious groups that work for nothing, and with volunteers who seek little more than enough to cover their expenses and insurance costs. Popular newspapers such as the *Daily Express* of the UK also criticized salaries strongly. In the mid-1970s UN salary levels attracted adverse comment in the US Congress; there were newspaper headlines that seventy-five UN posts carried salaries of $75,000 per year. The USA and USSR, the largest and second-largest contributors to UN finances, tended to join forces on financial issues. There was continuing pressure in the UN General Assembly to keep UN salaries down.

The view of the original United Nations when it was established was that its staff salaries should be high enough to attract the best-qualified people in the world. The combination of pressure in the General Assembly to keep nominal salaries down and erosion of the value of the dollar has resulted in a situation where some consulting firms, even in the UK, will no longer bid for FAO contracts. The governments of Germany and Japan pay salary supplements to their nationals to induce them to take up and remain in UN posts. Most senior American staff in FAO come to it only when they already have substantial pensions from their own government.

Supplementation of salaries from national sources is a negative development. It can mean that staff are still tied to their own government and are no longer independent internationally, as required under the oath taken when a professional joins FAO. A 5 per cent increase was approved from 1990.

PROMOTION PROSPECTS

While many people join FAO for a relatively short fixed-term assignment, its strength professionally depends on its maintaining a nucleus of competent, dedicated staff. This is not achieved by recruitment toward the end of a person's career. There has to be an inflow of younger staff who will absorb the experience of such an organization and grow inside it. Promotion from within is established FAO policy, backed up by a staff agreement that internal candidates will receive favourable consideration for a post that comes vacant if they are competent. As in most other organizations, promotion in FAO can either proceed rapidly, or take a very long time. This depends very much on both personality and luck. ADG Broadley said personality counted much more in FAO than qualifications. However, some of the aspects of personality he had in mind – gifts of attracting the attention of those above, of mobilizing support laterally, of seeing opportunities and pushing them at the right time, are the qualities of leadership. Luck is in being already in a unit which is raised in status during the course of some reorganization. When the big divisions of FAO were raised to departments in the 1960s the branch chiefs all became directors. Prospects in this direction may now be judged poor. Luck also comes in the form of the rapid departure of a succession of people immediately above one. Almost automatically the person in the post below will take over the post above in an acting capacity. He will still need luck, however, to be confirmed in it. This is especially true at the division director level, where the principle of maintaining an equitable balance of nationalities applies most strongly. There have been acting directors continuing as such for years, only to be passed over for someone brought in from outside.

With a nationality that is under-represented, promotion in FAO can be quite fast. H. Horning, a German, began as an assistant field adviser in Nepal and was appointed Director, Land and Water Division within a short span of years.

ADG Wells saw promotion in terms of the Peter principle, whereby each person rises to a level of responsibility slightly above his capacity. There his limitations become evident and he rises no further. In FAO, the way further is to become a country representative.

RISKS, BUT NOT HIGH

There are physical risks in working for FAO, but not significantly higher than in many other employments. FAO people fly long distances: a few have been lost in accidents. Again, the risk they run is normally much less than that of the man who flies regularly about Europe or the USA on business. A crash can be discouraging, of course: an FAO man who survived with four others in the tail of a plane that went down in the sea near Montego Bay cabled headquarters for authorization to continue his journey by surface.

In some places where FAO personnel live on field assignment, private home protection services are used. In one African country an FAO man newly arrived was inclined to quibble about the cost. 'Can I not call the police?' he said. 'Yes of course,' was the reply, 'but then you would have to go in your car to get them; they have no transport.' In relation to the number of FAO people working away from their homes complaints of robbery or similar illegal interruptions in the even tenor of living are few. Many maintain that the risks are much less in the developing countries than in their own. A man working in Ethiopia was either pushed, or had fallen, into a hole in the road. When he recovered from the shock he found that his pockets had been emptied. Always cautious in his statements, he said later, 'I still maintain that the people there are relatively honest.' I had my pocket picked while watching the King of Nepal perform a seasonal ceremony in Kathmandu; I have had it picked of a larger sum at a London airport.

There are risks to individuals from civil disturbance. Often they are more localized and confined to the people of the country than the media, with their interest in the dramatic, would suggest. Thousands were reported killed during the 1966 revolution in Zanzibar – a very small country. We had a man there helping start up an agricultural bank. When I asked him how it was during the revolution, he replied, 'We kept open,' he said, 'but there was not much business.' Once a model African country, Uganda became for a time a very dangerous place. I was visiting a project during a period of tension under Idi Amin. No one knew what to expect; there were periodic road blocks and the soldiers' reaction was unpredictable. One evening I was returning late with a locally based colleague. We came into the darkened streets of Kampala, and two figures with guns waved us to stop. 'Now we are in trouble,' said my colleague. 'Excuse me sir, do you know you are driving the wrong way down a one-way street?' was the soldier's remark, in a very polite tone. 'I am sorry' I said, turned round and drove back. Whilst in Zaire in the 1960s to help with food distribution, we were paid $20 per day extra as danger money. The only man who pointed a gun at me was one of the Tunisian soldiers there to protect us. He wanted the spare wheel of my UN Volkswagen because he had a flat tyre and his own spare was missing.

It is possible, of course, to be caught up in a government overturn unwittingly. This happened to a young marketing adviser in the

Caribbean, whose hobby was sailing. One free day, he says, he was approached at his yacht by a man who asked if he would please take him to a nearby island where there was a vet who could inoculate his dog against rabies. Looking favourably on an animal lover, our man duly made the journey and awaited his passenger for the return trip. He never appeared. Going back alone our man was immediately arrested – there had been a coup, and he had helped somebody escape. FAO got him out of prison, but he lost his job.

Health risks remain. While many of the inoculations recommended in the 1960s have since been abandoned, hepatitis is still common and malaria is more serious than before. Now, several types of prophylactic pills, taken simultaneously may be needed to combat anopheles which has become indifferent to some of them. FAO stomachs soon become adapted to the more common bacteria in food, but those of Kashmir and Nepal can incapacitate even the most hardened veterans.

FAO employment carries substantial insurance against health risks while on duty, with a disability pension at the worst. Some staff members have seen the primary concern of the FAO Medical Unit as to protect the organization against such claims. There was a man who had to be withdrawn from a post in Indonesia because of a heart condition. His doctor there had said he could not work, so he was brought back to Rome. We talked sympathetically about what he could do – perhaps retire to a small place in France? The next day I met him for lunch. He was angry; the FAO doctor had said there was nothing wrong with him. In that case perhaps we should order a good bottle of wine, I suggested. He looked at me scathingly. The next day, however, he was smiling. 'I'll buy the wine' he said. 'I have just remembered that when I left my doctor in Indonesia he wished me a good journey to Geneva. I never told him I was going to Geneva.'

Intestinal hazards may be slightly higher working with FAO than with some other UN agencies. Contact with farms and forests necessarily takes people into villages and leaves them more open to local hospitality. It is the meal at the house of the village headman or rural relatives of a counterpart that cannot for courtesy reasons be refused, that can put someone out of action for days to follow.

The hazards of hospitality can go further. An FAO man arriving towards evening in a village in Afghanistan was looking for an inn. A local gentleman saw him, and offered the hospitality of his home. This could not be refused. They went to a door where the FAO man was told to wait; his host would be a few minutes arranging for his women folk to leave a room free. Meanwhile, another local gentleman came up and said 'Friend, why are you waiting here in the street? Come home with me'. The FAO man explained about his invitation, but this was brushed aside. 'It is an insult to keep you waiting outside; come home with me.' He was a commanding personality and the FAO man felt constrained to follow him. This time he was immediately taken into the house. Shortly after the first would-be host arrived, banging at the door. 'Come outside, you unfaithful guest! Come outside and I will kill you.' His current host observed, 'Do not worry! If he kills my guests, I shall kill his guests!'

Hospitality can become a hazard in other ways. When Hungary became a member of FAO I participated in a mission to discuss the help it might need from FAO, and areas in which it had people qualified to be advisers in less developed countries. At each office we were offered a glass of barack (sharp apricot brandy) as an introduction to our discussions. After half an hour or so came another round 'to the good continuation of our discussion'; before our departure there was a third drink 'to the next time we meet'. Some FAO men who went to Russia told frightening stories of the vodka they had to consume to keep pace with their hosts.

PROFESSIONAL CONTINUITY

People in mid-career are sometimes reluctant to take up work with FAO because, they say, it will interrupt their career. They fear they will lose seniority, be out of the sight of their professional peers, and so be forgotten. This can be a short-sighted view. In most of its professional fields a stage with FAO will add to a person's professional experience both through acquaintance with differing conditions and through in-depth consideration of origins and relationships. Professors leaving university to work with FAO have found their new environment still sufficiently favourable, even stimulating, to write a further book, perhaps a better one. Experience in trying out accepted concepts and approaches under developing country conditions has been the basis of many theses and publications in professional journals.

This is not to condone approaching an FAO assignment with some advanced research methodology as the prime objective. Leaders of developing countries are correct in saying that to be host to such a professional is wasting their time. The vogue for econometric analysis has resulted in the application in various countries of techniques that could be judged from the start too elaborate for the quality of data available.

FAO has to watch for this. Its own reputation is at stake when it sends a professional to a country as its adviser. It also faces criticism when staff are seen attending academic meetings. Why are they not at their jobs helping the developing countries? Some of its administrators have been slow to recognize that presentation of a competent paper at a professional meeting adds to the stature of the author, and also to that of the organization with which he works. FAO is certainly restrictive in its financial support to staff attending professional meetings. Generally, however, it has been ready to agree that they could go on their own account, and incorporate the travel into their official programme so long as it did not involve any extra cost to the organization. As can be expectd, attendance at professional meetings held in Third World countries is looked on more favourably than those held in Europe or North America. As a last resort, a professional can obtain clearance to attend a meeting on leave at his own expense.

FAO does not go to the extent of the World Bank in keeping under wraps information assembled by its field missions. Its reports to governments are released for general use unless a government specifically requests otherwise. It does expect its staff not to use without clearance information supplied specifically in confidence or to quote individual countries to their obvious detriment.

There is an excellent library at FAO for the use of its staff and others. Most books on its subject area are there. It also provides access to still more comprehensive collections elsewhere via a photocopying service. Some FAO library services attract intensive use. Its reference room copies of the periodicals *Food Policy* and *Development Policy* are kept under lock and key; otherwise they disappear. There are waiting lists to read some of the new books. One suspects, however, that the immense stock of books in the stacks is tapped only lightly. I was considered a major client of the library: if I asked for books from the stacks twenty times in a year it was thought a lot. The riches of FAO for the professional are the technical reports coming in from its field projects, the current appraisals of reality on the ground, and the pearls of information and experience forthcoming from anecdotes over lunch.

In the 1970s the Rome international branch of the Society for International Development afforded FAO staff a chance to hear interesting speakers and debate current issues. Thereafter it declined. DG Saouma did not welcome independent discussion that might bear on FAO policies.

There have been many who worked in FAO for a number of years and then went back to a university with their professional reputation higher than before. One such was René Dudal. In FAO he led the preparation of a world soil map and rose to be the director of the Land and Water Division. Dudal kept his professional contacts and interests throughout his stay with FAO. He was President of the World Soil Science Association and received honorary degrees from several universities. He left FAO when he was still at his peak and took up a professorship at Louvain University.

Aziz El Sherbini maintained his professional standing in the USA and elsewhere throughout a similar twenty years with FAO and IFAD. He came to FAO from the University of Khartoum. During his time at FAO he kept up a flow of articles to academic journals, and when he finally decided to leave Rome he was offered university professorships in both Egypt and Jordan.

S. Holt, a biologist in Fisheries, came to FAO as a brilliant young scientist. He lost nothing by being in FAO. When he left he was writing an advanced scientific treatise.

Sartaz Aziz and Nural Islam both completed substantial books while in senior economic posts in FAO. Aziz, of Pakistan and Harvard, wrote in the late 1970s a book on China that has been read widely, though China has since moved away from his view. Aziz became a vice president of IFAD and then a minister in his own country. Nural Islam wrote about Bangladesh, from which he had taken refuge in Oxford before coming to FAO. He went on to a senior research post in Washington DC.

Another important point in FAO's favour for the professional is that substantial analyses and reports are issued with the author's name on them. Typically a study is attributed to FAO but the words 'Prepared by ...' appear on the title page. Articles contributed to external publications can carry the author's name specifically. It is then customary to state that 'FAO does not necessarily endorse the statements and conclusions of the paper; for these the author is solely responsible.' There was a struggle over this at the beginning. Sir John Boyd Orr wanted FAO staff to adopt the British Civil Service practice of issuing their writings unsigned. This is still followed in the Commodities Division in its analyses of market situations and prospects. Ralph Phillips claims that he was the first to convince his director general that if FAO professionals could not ascribe their names to their work it would not attract and retain the best people in its technical fields.

Presentation of a paper at a technical meeting, or publication of a book externally, requires FAO clearance. It is reasonably easily given. E. Zarcovich, a Yugoslav statistician, was denied clearance for a book maintaining that foreign advisers were wasteful, and that aid funds should be concentrated on training. This cut across FAO's current programme. It was also a questionable thesis. The Philippines, for example, benefited from large-scale domestically organized training on American lines, but has not developed faster than other countries starting at the same economic level. Zarcovich made hay out of it eventually. The book was published under the next DG with a foreword setting out that 'This is a book for which a previous FAO director general had denied publication.'

OUTSTANDING PERSONALITIES

The early leaders among FAO professionals were people who played a role in its establishment. D.B. Finn, first director of fisheries, and various others were members of national delegations to the conference through which FAO was founded. The International Institute of Agriculture was incorporated into FAO and its personnel came with it. These included Abensour, who became legal adviser to the director general, John Evans, for a long time a personality in the Commodities Division, Sakoff who followed events in Russia for ESP, and Moskovits who later made a hit as unpaid representative for Malta. In return for the diplomatic privileges this carried he secured for Malta a role as venue for meetings on Mediterranean agriculture far out of proportion to its size. These men carried on into FAO retirement at 65 as stipulated in their IIA contracts, and were around for a very long time.

Glesinger, Felsovanyi, and Fortunescu were together at the Forestry Institute of Vienna. Felsovanyi, it was said, had helped Glesinger escape to Sweden where he married and became a citizen. Glesinger fought hard for the establishment of a forestry division in FAO during its early stages in Washington. He then brought Felsovanyi in to work with him. Felsovanyi's family were originally bankers called Baruch. Ennobled by the Emperor Franz Joseph they adopted the Hungarian name of Felsovanyi and were famous in Vienna for their horses.

Some of FAO's most influential professionals during its first decades came to it from UNRRA. It was on funds transferred from UNRRA that FAO began its technical assistance. With them came staff that had worked on relief operations and the re-building of agriculture in war-devastated Europe, and also in China.

Rich in administrative experience was Frank Weisl. He was UNRRA's Deputy Director General and Chief Executive Officer. He had negotiated UNRRA agreements with the USSR, Poland, and

Hungary. When he came to FAO in 1946 as Director of Administration it had a staff of 114.

UNRRA also contributed some outstanding Australians – Walter Pawley, Al Faunce, and the Kesteven brothers. Pawley, according to a fellow Australian, literally clawed his way up in FAO. Another person under him saw in his cramped, barely legible handwriting a sign of brilliance. His great achievement was the initiation and management of FAO's work on its Indicative World Plan. Taking 1962 as a base, projections of food and related requirements were made to 1975 and 1985. Population growth alone would require a two-thirds increase in food supplies in the developing countries over twenty years, at existing nutrition levels and patterns of consumption. The combined effect of a larger population and higher incomes gave a projected increase in total demand for food in developing countries of over 140 per cent by 1985. This implied an annual rate of increase of 3.9 per cent. In most countries the domestic market was expected to take over from exports as the main driving force. To meet the needs of increasing city populations and of fertilizers, etc. for agriculture, the processing and other agro-industries would have to expand by 7 per cent. The plan stressed the need for a two-way movement of capital into industry as savings from agriculture and back into agriculture as investment for growth. Because populations were increasing so fast, rural incomes would rise only 1.3 per cent up to 1985. This set limits to the market it would provide for industrial products and showed the. need to increase rural incomes through supplementary activities such as public works and village industries.

These 'Proposals for an Indicative Plan for Agriculture' were discussed widely. The figures, of course, have been overtaken by events. In the late 1970s 'Agriculture in the year 2000' was prepared along similar lines. In the first of these studies FAO was finding its way: there was evident waste of manpower with periodic changes of gear over assumptions and procedure. Pawley parried comments on this with the Australian dictum, 'Two heads are better than one even if one of them is a sheep's head'.

Faunce was an agricultural engineer. He was a 'wizard', in the words of a contemporary, at putting together projects for UNDP, securing government support, and getting them under way. Mainly they were concerned with agricultural processing. He possessed a vivid range of Australian phraseology. His comment on one such project was that marketing stuck out 'like dogs' balls'. He lived elegantly with a French wife and on French food.

The brothers Kesteven also came to FAO from UNRRA in China.

Their forthright personalities were prominent in the 1960s – one as Director of the Animal Production and Health Division, the other as Chief of Fisheries Biology. They did not, however, set a very good example for FAO in population control. Together with other members of the family Kesteven they had twenty children.

Lena Passerini, one of the outstanding women of FAO, also came from UNRRA. She took a degree in agriculture at Cornell University at a time when few women were doing this, even in America, and then a degree at Bocconi University, Milan. With her European links and languages she was offered a job with the Société Belgo-Anglaise des Ferryboats. It operated the services between Harwich and Zeebrugge, and Harwich and Dunkirk. She then had a phase in Latin America on potato studies, polishing her Spanish. She came back to Europe when the allied armies liberated southern Italy, helping to provide seeds, fertilizers, and machinery for a resource-starved local agriculture. For this she was ranked a Major in the American army. Helping to bring the high-yield hybrid maize into Italy, and carrying out the rehabilitation of over forty-six experimental stations were two of her contributions. When the war ended Lena became a senior officer of UNRRA. When its European operation closed down and UNRRA transferred $5 million to FAO to complete its distribution of agricultural machinery, fertilizers, seed, etc., Lena was appointed FAO representative to the government of Italy. FAO was then in Washington. Lena had her office in the Villa Borghese, home of the old International Institute of Agriculture. When FAO moved to Rome it no longer needed a Representative to Italy, so Lena was offered a post in its Agriculture Division. There she developed FAO's approach to training courses, study tours, and technical meetings during its formative years. She was also a pioneer in the FAO Credit Union, as head of its credit committee. Many people owe a lot to her sympathetic approach. In Rome at that time, favourable sources of credit were sparse indeed. On retirement she continued to appear at FAO conferences, representing the International Council of Women which co-ordinated a number of non-governmental organizations with goals parallel to FAO. Her farm along the Pontina, south of Rome, became a contact centre for FAO staff and delegates to several score FAO meetings.

Mordecai Ezekiel was Deputy Director of the Economics Division when I first joined FAO. He came from a Jewish family in North Carolina which went back to the seventeenth century. He was much admired by people who had worked with him in the New Deal programme of President Roosevelt and well known to students of statistics for his path-breaking text on correlation analysis. Not too concerned with minor comforts – he once offered a breakfast guest coffee made with cold water – he would take the time to read my manuscripts and comment upon them. He saw his mission in FAO as the need to raise its standards in economic analysis. The staff refresher training programme he organized in the mid-1950s featured intensive courses by the American economists G.L. Mehren and C. Mitchell, and statistics courses by Professor Gini of Rome University. It was a course people still remember even though the Italians said Gini had the evil eye, and foresaw ill results in consequence. Professor Gini contributed his coefficient to land reform literature. Clyde Mitchell stayed on with FAO as a protagonist of socially-oriented planning in Latin America. George Mehren was the most impressive. He had experience in the administration of price controls and knew exactly what measures could be effective for different structures of production and distribution. He was appointed Assistant Secretary of Agriculture for the Kennedy administration.

Along with keeping up the quality of FAO economic and statistical work at headquarters, Ezekiel undertook major advisory missions in Greece and Poland. He was well received there because of his liberal political background. In Poland he made the point that farmers could still benefit from competition in marketing under a communist regime where they were served both by co-operative and municipal buyers of their products. This did not stand in his favour, however, in his home country. When Kennedy was elected President, Ezekiel was invited to be his adviser on foreign aid. There was a big farewell party in Rome; the Division bought him a carpet to go in his new office. Then came news that the Chicago *Tribune* had launched a campaign against him as politically unreliable. The offer was withdrawn.

R.O. Whyte of the Plant Production Division had a book on grasslands published in the UK. It was widely used in the universities. He left FAO when a man he had rejected for a P3 post was appointed over him as director of the division.

Al Aten led agricultural processing for many years. He was not a writer himself, but he commissioned technical manuals from a number of men whose experience he considered so valuable that it should not die with them. One was named Dowson, who wrote a definitive text on processing dates.

Dr Frank Parker was head of what became the Agriculture Department. He had done research on nitrogen fertilizers for Du Pont, the big chemicals firm. He came to FAO after many years with an

American technical assistance group working in India, helping to put into effect the first and second five-year development plan. He was impressed there with the impact of fertilizers on cereal output. It was under his leadership that FAO began its collaboration with the fertilizer industry which gave such practical weight to its training programmes and policy recommendations in this area. Parker was a co-ordinator and administrator with rich experience. He spoke thoughtfully, in clear and simple language.

Changing aeroplanes in Chicago I met a man I had known as a graduate student. He was then Dean of Agriculture at the University of Wisconsin. We talked of FAO. The only man there that had made an impression on him was 'that Greek.' This could only be Henry Ergas. He appeared in FAO in the 1960s as a protégé of Glesinger. He was to prepare and supervise projects to make semi-arid areas around the Mediterranean basin green again. He was a protagonist of integrated development planned from the top. Henry was much more than that, of course. Above all he was a master at getting close to people who mattered. He could ring up and talk on a first-name basis to more influential people in more governments around the world, one of his colleagues said, than anybody he had known. In FAO he charmed Orris Wells during a personally conducted tour of Turkey where he had lived when he was young. In 1964 he became head of the division funded jointly by FAO and the World Bank to prepare agricultural projects.

Henry became 'high priest' in FAO of the internal rate of return. Then he left. Some said that his highly personal style had tired people at the Bank. Another view was that he was wasting his time on an FAO salary. Doing the same thing for the Rothschilds would earn him a fat commission on every loan. He remained in Rome – financing the Italian government, people joked, until the taxes came in.

Viggo Andersen, popular director of Rural Institutions in the 1970s, was at Copenhagen when a shift in the Danish vote brought FAO to Rome. He had been DG Sen's candidate to head the World Food Programme when it was first established. His party piece was to sing in Gregorian chant spontaneous verses about his colleagues.

Brilliant in project organization and management with very limited funds was Chong Lee, a Korean who had taken a course at Harvard. During the 1970s he was FAO marketing adviser for the Asian countries, working from a regional office at Bangkok. He was adept at proposing training programmes adapted to local needs, convincing governments to sponsor them, and then finding the necessary funds. Often he would have more projects underway than half a dozen other men in his office, pulling in funds from America, UK, and other sources to supplement his own budget. He had the gift of making his participants do most of the work themselves, feel it was their thing, and ask always for more. With high FAO posts ahead he left to work in his own country and build another career there while still young enough to do so.

'Brilliant but erratic' was Orris Wells' summing up of Raymond Lloyd, a very exceptional professional in FAO. Lloyd originated the FAO Stamp Plan to be followed by the Money and Medals Programme. He persuaded heads of government to issue stamps featuring FAO and return to the organization part of the sales proceeds. Coins and medals featuring eminent personalities were sold in the same way. They reminded people of FAO's 'food for all' objective. Additionally, the funds accruing paid Lloyd and his staff their salaries and expenses. By 1980 there was a net profit of $2.8 million. This was available to fund more than forty small credit, women's development, and similar projects. Lloyd went directly to the people he wanted and succeeded. For his medals he sought out eminent women. They included Indhira Gandhi, Mother Teresa, Sophia Loren, and Farah Dhiba. The last of these was his greatest coup financially – the Shah bought all of her medals forthwith.

Soon after Saouma was elected Director General, Lloyd went into opposition. His operations were not making quite so much money as before, and there were some administrative problems. There was a move to put in a manager, which Raymond resented. Perhaps he was already looking for a new dramatic cause. He left FAO and single-handedly mounted a campaign against the Director General, using circular letters to governments and the local English language press. He attacked, especially, Saouma's changing of the FAO Constitution so that he could run for re-election after his six-year term.

Lloyd researched his case thoroughly and wrote well. If he had right on his side, it did not carry weight in a world of political realists. Lloyd continued his campaign on his own savings and at great cost to his personal life.

The most brilliant woman in FAO was Gerda Blau, Director for many years of the Commodities Division. She was already a Doctor of Law when she went to the London School of Economics in 1938 on a Rockefeller fellowship. She completed a Ph.D. in economics there, took up a research post with the Dominion Wool Marketing Organization, and acquired British nationality. F.L. McDougall found her for FAO. He was the representative in London of Australia, the major wool exporting country. They had many contacts. Impressed by her ability and interest in the practicalities of international co-operation he consulted her on a programme for fibres. Then she was asked to carry it out. In 1951 she was appointed Chief of the Commodities Branch, later to become a Division. She became the most well-known woman in international work, commanding the respect of government delegates, leaders in commerce, and of her colleagues.

Her final years with FAO were clouded and she had to abandon her directorship and work on her own. She maintained her standing in economic circles. Each year she played host to the distinguished English economists invited to the meeting of the Rome University Economics Society. At her cocktail parties I met Lord Kahn, Sir John Hicks, and many others.

Gerda Blau was a Jewess from Vienna. The woman who came closest to matching her stature in FAO was an Arab from Jerusalem. Her father, she told me, was head of the main post office there during the British mandate. Aida Eid came to FAO as a G3 secretary. She took a London University degree in economics as an external student and rose to be Deputy Director of the Investment Centre, preparing projects for financing by IFAD and the regional development banks. She was a systematic, competent professional with the strategic advantage of an ability to work in four languages, including Arabic.

These are the only women to have held director-level posts in FAO to date. During the 1970s there was pressure on the Director General to secure a more equitable balance between women and men at the higher levels. The word went round that a woman candidate should be found for vacant posts; they should be given special consideration. This did not come to much. The one lady appointed to a senior post had to be kept in the shadows. Then the Minister of Agriculture of Liberia, Florence Geneseth, came in view. At an African regional conference she was an impressive figure with a pile of gold cloth on her head as she spoke. She was given to understand that she might become head of the Agricultural Services Division. Then came revolution in Liberia; she walked over the Sierra Leone border dressed as a market woman. She came on to Rome, but now, the Director General did not want to know her; he had to work with the new government. Eventually she left for a study fellowship at an American university.

The low percentage of women in the higher posts of FAO is a reflection of male predominance in the fields it covers. The branch responsible for home economics and later women's development was generally headed by a woman. Ruth Finney, an exotic Haitian

American, was there in the 1970s. An inter-divisional committee was set up at this time to promote projects for women and ensure that their role received adequate treatment in other projects. She managed this smoothly and with confidence. Her husband looked spectacular, like Superman of the comic strips. One day they came back from a skiing holiday, he with a cast on his leg, she with a knee injury. This was slow to heal and she eventually went back to America.

Prominent during FAO's early years in Rome was Jean Fairley, Chief of Budget. She impressed male colleagues with her 'Boy Scout' approach to duty and Scoutmaster voice on the telephone. To the women under her she seemed a slave-driver who read books while they did the work. However, the staff had to be doubled and the chief's post upgraded after she left. Later, while in Pakistan with her husband, she wrote an authoritative book on the Indus River.

HEROES OF THE FIELD

The really great professionals of FAO are those who devote their working life to the developing countries, helping them directly. One of the strengths of FAO has been its team of experienced field workers ready to adapt to the conditions of one country then move on to another, and another. They build up a stock of experience more immediately applicable than any formal training; most of all they learn how to present ideas in such a way·that they seem to come from their local counterparts and so have better chances of implementation. Effective technical assistance is much more than studying a problem, applying some ready-made solutions, putting all this into a report, and handing it over to the appropriate minister. This is why FAO has preferred to build up and maintain its own field advisers rather than employ independent consulting firms. The best consulting firms are, however, also aware of the importance of continuity; they retain a core of permanent staff ready to spend substantial periods of time in a succession of countries.

Thousands of men and women have worked for FAO on field projects. It is difficult to do them justice. Even the editors of *Ceres*, the FAO periodical, were reluctant to feature their achievements. Exposure of starvation and injustice, great plans for the future – these were what held their international readership, not seemingly complacent accounts of success in raising crop or animal yields in one area of a particular country.

In the first years of FAO many of these field advisers were British or

Dutch. These were people who had retired from their own colonial services in early middle age and were still ready to work. Their experience was concrete and relevant. An FAO post for them was a logical continuation of their professional career. The Dutch had the advantage that they could also speak French. Then came an abundance of Indians, Pakistanis, and Egyptians. They also had the right experience and the requisite formal qualifications. For them, the FAO salary was manna from heaven. As one Pakistani said to me after his FAO field assignment; 'Thanks to this I shall be able to invite 1,000 people to my son's wedding.' He had already bought his black Mercedes.

French and Belgians with experience in Africa became available later. Finding good candidates for francophone Africa was always difficult for FAO. These countries wanted somebody new, but they must be French-speaking. Everybody turned to francophone Canada, but so few people there had any idea of French-speaking Africa. This was the problem also with another typical applicant for FAO field assignments – the American professor or extension man with sabbatical leave. He wanted a post for a pre-determined period; generally it would take him the same time to adapt to an environment that was completely different to the one he was used to.

The B.R. Sen prize was instituted to recognize men and women who had done exceptionally well on individual country assignments. It was awarded on the basis of recommendations by the technical units in FAO and the government of the country where the candidate worked.

Fred Scherer was the first to receive the prize for assistance in marketing. He had a great organizing sense as well as technical know-how and the gift to communicate it to the people with whom he was working. In the densely populated Kigezi Mountain of Uganda the 'white fathers' had taught local farmers to grow European vegetables. Climatic conditions were very favourable, but there was no market. Fred organized a co-operative system with over 4,000 members. They had their own wholesale outlets in Kampala. Trucks bringing supplies up to the mountains were loaded with vegetables for the return journey. To avoid glutting the market Fred timed the distribution of seeds and set weekly sales quotas. Later he trained indigenous Ugandans to run the maize mills and distribute flour after the Asians left.

Indeed Fred was so valuable to the country that there were difficulties in ending his assignment when he was ready to move on. They were compounded by an initiative of his wife, who was a skilled gardener. There was an occasion when Idi Amin was due to host a meeting of the Organization of African Unity. A hall had been constructed, but the ground around it still looked like a building site. Mrs Scherer offered to landscape it to the highest international standards in eight days if she could be given the resources. Amin gave her use of the army and access to the Entebbe botanical gardens, and the job was done.

Scherer was much praised also for his achievements in Brazil. He advised a government-sponsored market development organization, which set up rural assembly markets to attract buyers to areas where otherwise farmers would have to travel long distances to sell their produce. It helped small-scale retailers strengthen their operations and supply lower income consumers at more favourable prices. Fred Scherer was similarly appreciated in Nepal and Malawi. Just before he was appointed Chief of the Marketing and Credit Service in Rome, the Government of Malawi wanted him to take over direct management of the marketing organization which handled most of the grain sales in the country.

Contributions by FAO advisers in land and water management range from developing successful irrigation systems under very difficult conditions to training local advisers and managers over a large country. Harry Storer was a legend in the isolated Turkhana area of northern Kenya. He taught the people there to make the most of limited and precarious water supplies. When an official delegation came out to see the project, it was Storer who was held up in the air in appreciation, not the traditional leader or the minister.

Herbert Brammer won the B.R. Sen prize in 1981 for his work in Bangladesh. He started a detailed soil survey there, but after he had been there a number of years FAO moved him to another assignment in Zambia. The Government of Bangladesh asked for him back again to help identify areas of land suitable for planting with high-yielding rice and wheat. His training of young people for extension work was another major contribution. When he left, 15 to 20 per cent of the Bangladesh rice crop was from high-yielding varieties. He played a leading role in this achievement. Brammer said that the substantial time he was able to stay in that country enabled him to see the results of his work.

The approaches adopted by men taking up advisory assignments have differed sharply. At one time we had two specialists on rice marketing in Burma, a German with a good knowledge of world markets and export procedure, and an Indian who was developing quality controls in buying from the farmer. The German made every effort to adapt to the Burmese way of thinking. He spent Christmas in a cave meditating Buddhist-style. The Indian spoke to his counterparts with an arrogance that made me wince. Yet it was the Indian who was asked to stay on for another year.

To help make sure that field experience is carried into FAO management and to ensure that headquarters is kept conscious of field project support needs it is FAO policy to look first to field staff in filling headquarters vacancies. G. Bonte-Friedheim, Assistant Director General for Agriculture in the 1980s, began his international career as an adviser to the Government of Kenya. P. Newhouse, in charge of early warning in Commodities Division, followed the same route up. He worked in Africa as a marketing adviser, then as a project manager. Scaillet, Director of Agricultural Operations in Rome, came in from assignments in Peru and Brazil.

Many field advisers prefer to work directly with a government department, seeing an intermediate-level post in Rome as unsatisfying. Some develop such comfortable working and living arrangements that FAO has had to set limits on the duration of their assignment.

Governments appreciative of the work done by an external adviser often request a continuation year after year, rather than make an effort to ensure that their own people are trained to take over. In many places, however, the contrast between the comfortable life at FAO head-quarters and the conditions under which field advisers work is sharp indeed. It was a visit to an FAO team on the Kaedi project in Mauritania, with living conditions extremely arduous, that stirred Director General Saouma's impatience at some of the attitudes he encountered with staff in Rome.

PICTURESQUE

Positioned on the Mediterranean, international in character and embracing a wide range of support as well as technical fields, FAO has attracted to its staff a good share of the picturesque. They could be bizarre and original by the background from which they came, the personalities they presented, or the activities they pursued under FAO cover.

Nikolich and Kallay were old eminences. Nikolich had been an ambassador for Yugoslavia. In FAO he started as a messenger. Kallay was the son of the Prime Minister of Hungary under Regent Horthy. He managed the European Association for the Control of Epizootic Animal Diseases. Ignatiev was old-regime Russian and immensely tall. His father was Minister for Education for the last of the tsars. Ignatiev prepared the order of the day for FAO's inaugural conference at Quebec in 1945 and was the second man to be put under contract by it. He became head of the Soils Branch. Incongruous in a Scots kilt for ceremonial occasions, he was one of FAO's best-known figures.

Edward Szczepanik worked quietly and said little. He came to FAO from a university in Hong Kong. When we were working on the Indicative World Plan he attempted a correlation between levels of economic development and efficiency in marketing. It never came to publication because of uncertainty as to which were the causal factors. Years later I talked to a woman on the cross-Channel boat to Newhaven. She said her neighbour worked in FAO. 'You must know him,' she said, 'a funny man who does not speak to anybody.' Shortly after that I saw an announcement that Szczepanik had been acclaimed head of the Polish government in exile. There was a picture of him wearing the traditional gold chain of office over a white silk shirt.

Nigel Heseltine had the air of a swashbuckler. When I first met him he was about to make a journey across the Sahara from Tripoli to Chad, and he asked me to go with him. He was already an author – of a book titled *The Mysterious Pregnancy*, published in Dublin. He would write a new book about the desert journey; I should take photographs and push the jeep if it went down in sand. Friends advised me to call it off. He was a quarrelsome man; he could well drive off and leave me in the Sahara. Later I heard that he took with him a Frenchman; they only reached Tripoli before they parted company. Heseltine went on alone. He had with him, he said, 100 spare parts, but not, in the event, the one that was eventually needed. The jeep remained in the desert, and he continued by hitch-hiking and published his book. When I asked later if I could see it he replied predictably, 'Buy one.' We were together in Zaire at the time of the UN presence. He was found to be carrying a gun, which was against UN rules, and he had to go. Next he was in Madagascar as a planning adviser.

R. Aubrac came into FAO quietly, to work with Glesinger on integrated development projects. Only later did we know he was a leader in the French Resistance during World War II. A contemporary he suspected of collaboration and treachery published a book recounting his version. Aubrac felt constrained to issue a rebuttal. Imprisoned with all his group, except one, he was saved from certain death by his girl friend claiming that she was pregnant. Her plea to his gaolers that he should be allowed out under guard for a brief marriage ceremony was heard. He was rescued *en route* to the wedding. His colleagues in prison were never seen again. These events came up in the

French press in the early 1980s. A second notable attribute of R. Aubrac was that at University, he had shared a room with Ho Chi Minh. He was one of very few westerners who could obtain direct access to him during the Vietnam war. Periodically during the build-up for the Nixon/Kissinger negotiations he left Rome on unexplained travel, seeking a basis for a settlement.

Earna Bennett, who led the General Service strike in 1973, was reputed to have been parachuted behind the lines to work with Greek communists during the Civil War. This earned her considerable standing with FAO staff. Later it appeared that she was then Ernest Bennett, who later married and was father of two children. Before joining FAO's Plant Production Division she had undergone a sex change. This was revealed by a consultant who knew her in previous professional circles. In 1990 she lived at Bracciano with a woman and was barred from entry to FAO premises for reasons unstated.

John Cairncross came into the Economic Analysis Division as the brother of a well-known economist and a man who had been selected for the élite administrative class of the UK Civil Service. He was to help with report-writing in English. Later it came out that he had been the 'fifth man' in the Maclean–Burgess spy scandal. They met at Cambridge in the 1930s and continued to transmit information to Russia when World War II was over and the West and East had moved into a cold war relationship. FAO was only 'bread and butter' for Cairncross – along with the handfuls of biscuits he was seen to stuff into his pockets at divisional parties. His primary interest was in writing, including a fascinating Penguin publication on polygamy. He attracted further publicity when he was arrested for trying to carry over the border into Switzerland a suitcase containing 52 million lira. This contravened Italian exchange controls. Various public figures (including, it was said, Graham Greene) were enlisted to help convince the Italian *Guardia di Finanza* that John Cairncross was an absent-minded literary figure who had not realized that Chiasso was in Switzerland. His appointment to a professional post in FAO in the first place also had its implications. The director responsible said to me rather defensively, 'If he passed the British administrative Civil Service examination he must be good.' Why he left that service with its comfortable salaries was an obvious question mark. In fact, he had been obliged to resign following investigations. Later, I remembered that at an agricultural economists' association meeting in Minsk I had been told that this director was a card-carrying communist.

Most open of the FAO spies was Jack Ward, an American who

managed field projects when I first joined. I admired him for his stock of witty stories. He left after a few years to join a military intelligence school as an instructor. Later he was in Brussels with the Soybean Promotion Council, well known as CIA cover. Then he died suddenly in Geneva – home of the café Clemence where spies sold their secrets during World War II.

After the Cairncross publicity we saw spies in every office. The sugar specialist in Commodities Division who travelled all the time without need to account for himself – he must be funded by the CIA. My maid thought I was a spy. There were attempts to enter my apartment; once a man was seen at my door wearing the classic raincoat. 'He was middle-aged,' I was told, 'not the type for a thief; they must be looking for documents.' Then one day I was telephoned at FAO – my apartment had been blown up! My maid thought it was a bomb in the refrigerator for me.

5

GENERAL SERVICE

As the base for a programme that is near world-wide, FAO headquarters has a large supporting staff. However committed an idealist may be to the cause of international food and agricultural well-being, he won't work for long unless he gets paid and is remunerated for his travel and accommodation. The efforts of a specialist designing a processing plant will be lost if his plans and instructions are not translated into the languages of potential users, reproduced, and distributed. An effective advisory and development agency must have its pay unit, its documents clerks, its translators, and interpreters. Secretaries are needed to handle the flow of correspondence, purchasing units to acquire equipment, dispatch clerks to distribute materials.

It is this face of the UN agencies that has led them to be attacked as bureaucracies living on the backs of the Third World countries. Their administrative and support staff are in full view. Their constructive work is spread over many countries and is less easily seen. Many of the support jobs are repetitious and. dull. The people who do them can hardly be expected to behave as ambassadors of their organization year after year. The coffee break, who they will meet at lunch, what they will do in the evening or at the weekend – these are the parts of their lives that are important for many of the younger staff. Depending on their cultural background, some of the older ones think a lot about the privileges and allowances that go with an international appointment. Any large organization, even one based on voluntary contributions, has to live with these human attitudes. Considerable administrative skill, and periodic phases of financial constraint, are needed to make the best of them.

The general service in FAO ranges from messengers and guards through secretaries and clerks up to personal assistant to ADGs, who handle highly confidential issues and correspondence. Carpenters and plumbers are paid for competence in their profession. Assistants to ADGs are paid much more because of the prestige of the person they assist, and the responsibility of the work they do.

In 1981 general service posts at FAO headquarters numbered 2,493; additionally there were 216 with the regional offices and joint divisions, 294 in country offices, and 679 linked with field projects. This makes a total of 3,682. By 1987, financial stringency had brought down the number of general service posts, taking into account all sources of funds, to just over 3,000.

POSTS, GRADES, SALARIES

General service posts are graded from G1 to G7. A G1 monthly salary is around $1,400, increasing by annual steps. There are G1 posts for manual workers, messengers engaged in sorting and distributing mail, in handling and packaging publications. Guards are generally graded G2 and up. The G2 salary is around $1,500.

Clerks, graded G2 to G5 according to the responsibilities involved, are employed in administrative offices and divisional registries, in the library and in divisions involved in collecting and analysing statistics and economic information. The Commodities, Economic Policy, and Statistics Divisions in the Economic and Social Department make considerable use of clerks. At one time ESP Division had two administrative clerks graded G5, a research clerk G4, two clerks in the registry G3 and G4, and a junior G2 clerk. The highest general service post in this division was a G6 administrative assistant who was concerned with the recruitment of staff for headquarters and field posts, their remuneration, leave, etc. These general service posts do not require degrees, though people with degrees may apply for them as a way of entering FAO. Clerks in FAO tend very largely to be locally-recruited. There are internal training programmes that help staff joining at a low level prepare themselves for more responsible work.

An important general service unit was the internal printing shop. Highest priority went to the preparation of documents for conferences and meetings. This material had to be out in time; otherwise FAO faced a flurry of complaints from delegates and participants. So it maintained its own printing unit with the latest offset equipment and established a strict procedure for the preparation and clearance of material for reproduction. In an emergency these could be delivered at 5pm and be ready for distribution in 100 copies the following morning. It was under the Publications Director, later ADG, Mandefield, the FAO internal printing attained this level of efficiency. Offset documents could be ready a few hours after the originals were received. There was an extra cost in overtime payments. Cynics said that the printing unit went slowly during normal hours so that more work would be done on overtime, and that this was tolerated as part of an understanding that documents would always be ready on time when needed.

The printing shop also turned out large quantities of technical bulletins, periodicals, and other FAO materials in a range of languages. Books published by FAO were handled by external commercial printers under a system of competitive bidding. Generally this was won by firms based in Naples. In spite of its higher labour costs, a firm in the Netherlands once made the best offer. FAO decided, however, to continue with its local firms because they were more accessible for last-minute changes, proof checking, and the discussion of illustrations.

While limited in size, the gardens round the FAO buildings have always been very well kept. The lawns maintain billiard-table perfection – no mean feat under a Mediterranean climate where, as someone once wrote 'lawns are for Onassis'. Flowering plants bloom successively throughout the year, and pot plants are maintained in a hidden greenhouse to be brought out to grace major conferences and receptions. The three gardeners responsible – one with a diploma – are also general service.

FAO has been able to economize on secretaries with general use of the word processor. This eliminated extensive re-typing of corrected drafts. Previously, and for most people still, the secretaries in FAO were the heart of its general service. In each technical division there would be some thirty secretaries supporting forty or more full-time professionals. Most of the secretaries were girls, with a dilution of men from Sri Lanka and elsewhere in recent years. The basic starting grade for a secretary is G2, that of a mono-lingual typist with a salary of around $1,500 per month. Monolingual shorthand typists or stenographers and bilingual typists (G3) earn $1,600. G4, bilingual stenographers, earn $1,700. Secretaries occupying posts that do not call for competence in additional working languages can obtain a language allowance on passing an FAO test. Secretarial posts G5 and up tend to follow the status of the office with which they are linked. Thus service chiefs rate G5 secretaries, division directors G6. At this level the amount of typing involved declines – unless the professional concerned likes to write a lot of letters and papers. The higher status reflects the responsibility associated with the post, the confidentiality of some of the materials handled, and the need for discretion over staff and visitors seeking the boss's attention and support. These higher posts also involve some supervision of the other secretaries in a unit, seeing that work is distributed equitably, and completed correctly and on time.

In November 1977 the FAO staff magazine featured 'The vanishing stenographer'. Recruitment policy was to engage general service staff locally. Advertisements were placed in the Rome daily, *Il Messagero*, and the international edition of the *Herald Tribune*. Response was poor. The salaries offered were comparable to those paid by big international firms, continuity prospects were good; Rome was an attractive place – why could so few people be recruited? The conclusion was that FAO standards were too high. Only 20 per cent of the local applicants passed its test. Secretaries could obtain posts typing only 35 words a minute – the main requirement of them was that they handle correspondence somehow, answer the telephone, and be pleasant to clients. This was found to be the case even in England in the 1970s. FAO had to drop its requirements of 50 typed words per minute and 90 words per minute of shorthand.

In fact, relatively few professionals in FAO were in a position to dictate at high speeds. For many of then, the FAO working language is a second language; they must check dates, facts, and quotations as they go. The Sir Robert Jacksons who dictated their 'easy' letters to a secretary in the car taking them to work are relatively scarce in the UN system. Many professionals make drafts in long hand, counting on the secretary to smooth the language and check the spelling. Some enthusiasts use a dictaphone to put material on record when they are in the mood, to be typed up later; but in few offices is this continuing practice. G7 posts carry special responsibilities in assisting a senior officer. Initially very few were established. They tend to be established for 'human' reasons. The occupants of G7 posts have often been people for whom the post or grade was established *ad personam*. It should not necessarily continue after they depart. Many were ladies who had been personal assistants or senior secretaries to director generals and assistant DGs who had retired. Their successors could not face day to day comparison with the abilities and experience of the previous holder of the post. They preferred to bring with them a secretary with whom they already had a good working relationship or to pick somebody new. To make way for this the current incumbent would be pushed upstairs to a post created for the purpose. Her preoccupation then would be to find something interesting to do. In some case G7s were created to reward an assistant whose work was appreciated, but who lacked the formal qualifications for appointment as a professional.

General service salaries in FAO are set at a level comparable to those paid for posts with similar requirements and responsibilities by the best employers in Rome. This is designed to enable FAO to draw on the best-qualified people available in Rome both local and foreign. For many years after its arrival, FAO was considered a very desirable employer. Great efforts were made by parents of children coming out of high school and university to obtain posts in FAO for them. In addition to the salary there was continuity of employment, health insurance, and pension rights.

Once safely inside FAO many of the general staff tended to seek additional benefits. They wanted a liquidation payment equivalent to one month's salary per year of work, as was customary in Italian employment, and access to the FAO commissary intended to help foreign staff who could not find in Rome products to which they were accustomed in their own countries. A general service staff union was formed, the Unione Sindicale, to protect and advance their interests.

A build-up of staff feeling against employers formerly considered very favourably was widespread in the late 1960s. During the early 1970s it reached the point of full-scale strikes in a number of UN agencies. In Geneva, UN staff were out for several days. They won salary increases that were later considered excessive. Cost of living increases were then frozen by instruction of the UN General Assembly until a substantial part of the increases won by the strike were absorbed. In FAO the strike was short; banners of protest were displayed, and general service staff left their offices to parade in groups around the FAO buildings. Concerned at the adverse publicity and the impact of such scenes on visitors from member governments, FAO's management gave way quickly. Most of the outstanding issues were settled along the lines sought by the Joint Action Committee of the strikers. FAO returned to its normal amicable working relations.

Issues between the strike leaders in FAO and its management included staff claims for:

1   adequate machinery to represent staff views in management decision making;
2   the right of staff representatives to devote part of their working time to representational functions;
3   conversion to continuing employment of staff who had been employed for a number of years on short-term appointments;
4   adequate adjustment of salaries in the face of inflationary conditions.

General staff salaries were set on a new base, with provision for increases on a par with those achieved by the best-paying comparable employers in Rome. In relation to the work done, the resultant salaries were considered high by many who came into FAO from commercial employment. Staff representatives countered this with the argument that the organization had systematically downgraded posts in compensation. A G3 had become standard for secretarial posts, whereas formerly many were G4. New incumbents started at the G3 level even if they were experienced and had held higher grades elsewhere. The requirement that anyone employed on a short-term basis for several years must be given a continuing appointment restricted flexibility. The cost to the organization of keeping on until retirement some of its less useful employees whose appointments had to be made continuing has been considerable. A limitation on the impact of the concessions made as a result of the strike was that they applied to existing staff, not to personnel recruited subsequently.

To retain flexibility of expenditure, the proportion of the FAO budget taken up by continuing staff salaries was brought down to 60 per cent or less. This required a much-increased use of consultants and the contracting out of work on a piece as opposed to time basis. It could mean higher costs per person, but not necessarily higher costs for the services performed, if the consultants and contractors were familiar with FAO procedures and requirements. In practice, however, many units accumulated studies that served only to pad out their shelves.

Long-standing defenders of general service staff rights in FAO have been Norah Connolly, T. Rivetta and G.W. Dodi. In 1974 they received powerful support from Earna Bennett, a professional in Plant Production Division. She had a strong voice and cutting phraseology; 'If the Director General had a twitter of sense he would ...'. Bennett gave way on the point that the gains of the strike should not necessarily apply to staff recruited subsequently. Her motives were never clear. She was given, it was said, a one-year salary until she became 55 and became entitled to a pension – on condition that she did not enter the FAO building.

LOCALS AND NON-LOCALS

FAO operates in a country where none of its main working languages (English, French, and Spanish) are indigenous. While some Italians knew these languages fairly well, they tended to be people who had travelled or were in the tourist business, not potential candidates for general service posts in FAO.

The organization had to offer special terms to attract the staff it needed. Travel home on leave would be paid every second year. There were allowances to help meet the cost of settling in another country and additional salary for a second working language. The elite were those who came with the organization from Washington. Another group came into FAO from a hectic life with the Italian film industry. Kathleen Edmondson and Cicely Cheyne spent 30 years with FAO after helping Robert Wise make 'Helen of Troy'. O'Riordan, Swanson, and Rossiter came in after 'Quo Vadis'!

These non-locals, as they were known, continued to be key staff years later, even when the number of suitable people recruited from Rome was larger. Especially valuable were the older English, French, and Spanish mother tongue secretaries who had been with FAO for a long time. They could advise on procedures not only to more recently recruited office staff, but also to many of the newer professionals. Newcomers at the highest levels were particularly dependent on these senior staff. They became very expensive. If they had been brought by FAO from another country they had the right to paid home leave every two years. Some who came with FAO from Washington had home leave rights to North America and Australia. A large number came from the UK and Ireland. Their home leave rights became conspicuously expensive if they had to cover also a dependent spouse and children. FAO tried to phase out these expensive office staff. Selected Italian girls known as cadets were recruited and trained specifically for work in FAO. Many went on to senior posts. FAO could also count on a steady flow of English mother tongue girls who would come to Italy on holiday, or to be with friends, and then want to stay.

Champion of the non-locals in defending their rights was Norah Connolly. with her leadership they established their own staff association distinct from the Unione Sindicale which they felt to be jealous of their privileges. Connolly had been an officer in the Women's Naval Service during World War II, organizing naval support to merchant convoys across the Atlantic. She took up staff representation out of a strong sense of justice and relish for a struggle. I asked her what was wrong about FAO's style of management that led to protests, appeals, and a full-scale general strike, when to outsiders FAO seemed to offer very attractive terms of employment. She contrasted it sharply with that of the organizations where she had worked before; and with the thinking of local staff influenced by rather elaborate and egalitarian Italian labour legislation. Moreover, while FAO talked of promotion by merit, there were always people who had been pushed by their governments.

There was also Mr Weisl, in charge of FAO's administration from 1948 to 1967. He had picked up the American style of 'hire and fire'. He had a sense of rough justice but it was to be his justice, not that of his employees. FAO had also been inclined to respond to its needs at a particular time without thinking through the implication for the future. When it needed English mother tongue secretaries to be able to carry on, it brought them in from the UK and Ireland. It was not foreseen that they would marry, have children, and expect the same comprehensive home leave and educational privileges as male staff recruited from those countries. It did not anticipate that girls recruited 'locally' while in Rome to be with a boyfriend would see it as unfair that they should forego privileges accorded to colleagues of equal qualification who happened to answer an FAO advertisement in their own country. Norah Connolly retired in 1981. General service staff in FAO will be very lucky if they find again someone to take up their cause with the same courage and ability.

LIFE IN THE OFFICE

It was difficult in FAO to maintain an even distribution of work among the secretarial support staff. They were allocated to professional units according to standard administrative criteria. One secretary for the unit chief, another to support two assistant professionals, and so on. This took little account of varying work output and the demands of the professionals concerned. One professional would turn up casually and begin the day catching up on the news. Another would arrive early, have a dozen letters ready for typing, and fume because his secretary had been held up by traffic. There was also the situation where a secretary worked for two or more professionals. One would have urgent letters to be typed; the other would insist that his report had priority.

A typical working day for some secretaries would begin with arrival at the office within ten or twenty minutes of the official time – pretty good, considering. She would look into her office and leave her coat to make sure people knew she was there, then go for a coffee and brioche at the bar. This would be a likely place to contact her if necessary. Then she would really get down to it and dash off some letters to have something to show before morning coffee. If there was a long report to type this could be a drag, but then lunch was not far away. Some days she might have to skip lunch and take a bus into the centre of Rome to do some shopping, or get to an electricity, telephone, or municipal office that was only open to the public between 8.30am and 12.30pm. But there would be a chance to get a sandwich some time during the afternoon. A conscientious girl in this situation would make a point of remaining in her office at least a few minutes after everybody else had gone so that she could say she stayed late to make up.

Another secretary would be there early, already worried. Her boss had a meeting coming up. The papers had to go for processing in a couple of days. It would be slog, slog all day. The paper would then go to another unit for checking and come back with various changes. This could mean re-typing numerous pages.

Many women committed themselves entirely to their FAO jobs. For one such woman it was her first real job. She saw herself as helping to raise levels of living for the less advantaged people of the world. She was assigned duties that were all engaging, but acceptable because of the responsibility they implied. In the morning she rose before 7am to be sure of being at FAO at 8.30am though she lived only a few kilometres away. Child and perhaps also husband were put second to FAO. She was angry about staff who she felt exploited the organization, sharply critical of those who came in late or took leave on false pretext. She was angry too when additional posts were established to take on some of her responsibilities. The organization was much larger and its finances were more complicated, but she saw this as unnecessary over-staffing. She did not travel far on leave or at the weekends. She went home early on a night out; she wanted to be fit for FAO when it opened on Monday. When she retired she had nothing to do; FAO had been her life.

To some FAO staff it has been a concern that most FAO 'bosses' call their secretaries by a first name, but are themselves addressed as 'Mr ...'. 'Real people,' it was claimed, would address each other by first names irrespective of status; they would then work better together as a team. FAO brings together people from many different cultures. There were many bosses who attracted the firm loyalty of their secretaries without such 'equality'. Does one, indeed, conclude 'what's in a name?'

It was once proposed that there should be a 'Miss FAO'. This would add interest to the lives of people doing repetitive jobs and bring prestige to the winner. She was chosen at a summer ▪barbecue for personality, interests, and achievements. There were no subsequent contests – the first winner refused to return the trophy.

LIFE OUTSIDE

Once the novelty had worn off, many general service staff saw their work as a chore to be got through as lightly as feasible. Their real interests were elsewhere. There were men who lived for their favourite sport, who waited for chess in the staff co-operative lounge at the end of the day. Women pursued a range of interests in parallel with their office duties. Some cared for the cats which adopted the FAO parking area, bringing food each morning, arranging for veterinary treatments. A kind person who picked up one of these cats 'to give it a good home' had to bring it back smartly in face of protests by its self-appointed godmother. There were the artists led by Clara Hemsted. They had annual shows where they bought each others' paintings. Some were very good.

Others had business on the side – such as wearing striking clothes discounted by a boutique that could be re-sold at a profit. Renting rooms to visitors would be another such activity. Access to an FAO telephone number was a great advantage. Coupled with normal social usage, the pressure on the FAO telephone service led its management first to exhort a more sparing approach, then to cut sharply the number of outside lines. External calls might then have to wait until 'the boss' was away and his line could be used without need for explanation.

One of the great general service characters was Olive Wherett. She was an outstandingly good secretary. She typed correctly at exceptional speed and was always ready for overtime. She was known best, however, for her passionate interest in cars. She liked them fast and of vintage quality, like the Alfa Romeo Freccia d'Oro of the late 1940s. She was a purist for performance, changing the plugs on her Ferrari for the open road when it had warmed up getting out of Rome. She asked me to be her co-driver in the Mille Miglia – the famous annual race round the 'consular' roads of Central Italy. Friends talked me out of it. It was stopped in the 1960s when a Spanish marchese killed thirteen people. Olive was a committed vegetarian; she ate regularly at the expensive Apuleius restaurant near FAO, but only pasta. She would tolerate my company for an occasional meal – provided I also kept off meat. To support her cars, it was thought, Olive lived very economically – in Ostia when it was out of season, in a small hotel room in Rome during the summer. Sometimes she slept in her car. There is also a story of an FAO guard checking the darkened corridors of Building B in the early morning hours. He fled in fright at the apparition of a ghost-like figure in flowing night clothes. It was Olive heading for the loo from an office where she was spending the night. Olive died in the late 1970s. She was buried at the Prima Porta cemetery in the part where graves are opened and closed by bulldozer. A colleague at FAO conducted a service according to the rites of the Protestant sect to which Olive belonged. Later, it was found that, in addition to seven cars parked in streets nearby, she also owned four apartments. None were rented out; she slept in each one by turn.

There was also a beautiful general service spy. She came to FAO in retirement, as it were. She had married a man of another nationality, which could bring a conflict of loyalty, so she needed another job. It may be pure coincidence, but once when I was leaving on duty travel she asked if I could bring back maps of a disputed frontier area; they could not be obtained in Europe. Later I heard that she had gone there on a walking tour. We also had a Russian with strikingly clear blue eyes. She told me that the KGB let her leave with a Kenyan student husband on the condition she would keep him on the revolutionary track. She went back to Russia frequently, sometimes leaving her children with her mother there. In Fisheries Statistics she was well placed to transmit useful information both ways.

MOVING UP

With the starting salary for each grade in FAO went a series of within-grade increases. These came automatically after each year or two years as specified – subject to confirmation of satisfactory performance by the post supervisor. For most people this signature came without issue. There were others for whom 'getting that signature' was a matter of lively concern. For weeks before it was due they would make a point of coming in to work early, being helpful and charming to their boss. For the supervisor not to sign called for considerable determination. It would mean continuing tension between people in day-to-day working contact. It would attract an enquiry from Personnel – 'Why was there no signature this year? Last year there had been no complaint.' If the supervisor held his or her ground there would be a request for a transfer on the ground that he or she was 'difficult'. Movement out of such situations was best for all concerned. The Marketing group had two secretaries that we thought unsatisfactory. They were often away, or could not be found, worked slowly, and on occasion were offensively unhelpful. They were said to have political 'attitudes'. Transferred to other units they were obviously happier.

The established path to promotion is by applying for a higher grade post when one becomes vacant. This will be in competition with other candidates. The unit concerned is under obligation to consider all applications and have eligible candidates interviewed. In practice, it is likely to prefer someone who is known, often a person who already works there and merits promotion. In this case the vacancy announcement may indicate that 'an internal candidate is under consideration' and the competition will be nominal. Where there is no internal candidate and applicants are few, then the boss's reputation has gone before him. 'Oh you poor thing!' was the greeting to one new girl proud of her FAO appointment.

The big jump is from general service to professional. How to surmount this obstacle was much discussed in the 1960s. Broadly, the criteria for entry to professional ranks was possession of a university degree or equivalent formal qualification. This was thought to keep out many people of ability far exceeding that of some who had scraped through a university. This requirement could, however, be waived, notably on ground of demonstrated ability to carry out the duties of the specific post at issue, i.e. evidence of performance in practice. Many general service staff in FAO were able to move to professional posts on this basis, particularly in administration where knowledge of procedures counts for a lot.

Examples of FAO staff who began with lowly grades in G and rose to high professional posts are much-quoted. A. Leeks, who started as a G4 clerk and became Director of Commodities Division, is the most striking example. K. Abercombie, who retired as a D1, is another. Many UK people with degrees went to FAO for general service posts in the 1950s because its goals appealed and career opportunities in their own country were limited.

Franco Granieri showed how one could rise in FAO through skill in administration. He entered FAO directly from the Italian air force as a lowly G and ended up as a D1 administrative manager for the FAO unit preparing projects for the World Bank. While others saw problems in administration and discussed them, he came up with solutions. Another who went the same way up to D1 was Pokorny; his great asset, somebody said, was a glowing smile.

There are many others who began modestly and rose to considerable heights on their own competence and initiative; Jacqueline Cane was one. She began as a secretary. When asked to take on some additional job she was ready – 'If I am to be here from 8.30 in the morning until 5.30 I may as well work.' After a phase in Zaire in the FAO representative's office she moved with him to the FAO Conference Unit. This put her in a position to be helpful to the Chinese when they rejoined FAO. From this came a study tour of China for FAO personnel that was well-managed and informative. Jacqueline retired from FAO as a middle-level professional and set up in Capri. As a focal point for

English-speaking visitors she had been filling a gap left open since the death of Gracie Fields.

Lily Pustina started in FAO as a cadet in training. By the 1980s she was a P4 administrative officer. Marcucci made little headway as a G in Commodities; in Administration he rose smartly to D level.

One can also move up by moving out – to come back in later at a much higher level. My first secretary in FAO, Marianne Boehlen, is the ultimate example. She was Dutch and spoke a handful of languages. I valued her, but she had her failings. 'Buffer stocks' once appeared as 'bugger stocks' on a technical paper sent for processing. She first transferred to the United Nations in New York, then because of her knowledge of languages she became the confidential secretary of the Chancellor of Columbia University. After this came a post with Lady Barbara Ward, a leading Catholic intellectual. There, Marianne said, she was telephoning the Vatican weekly. She took a degree at an American university while in these jobs and moved on to a professional post in West Irian. She came back into FAO as an administrative support officer with a UNDP project in Burkina Faso and thence to Rome as a P3. When I last saw her she was Deputy Director of UNITAR, the United Nations Institute for Training and Research.

In recent years the G or P issue has declined in significance. Staff union pressure in the 1970s put general service salaries on such a favourable basis that they overlapped those of professionals. FAO general service salaries were linked to the highest prevailing level with the best Roman employers. These went up with the Italian economic miracle and jumped again in the 1980s. Professional salaries came under pressure in the United Nations General Assembly in the 1970s; in European currency terms they went down again with the dollar in the mid 1980s. A G7 came to be better off than the P2. While there were a few who welcomed promotion to P2 or P3 in their later years with FAO, as a prestige recognition of their service, others coolly held on to their G6 or G7, foreseeing a more favourable base for their pension.

Moving up via a general service post in FAO is not just a matter of within-grade increases and promotion. One of the richest prospects for a woman is to marry her boss – or someone else's boss. The dream of the Italian girls who came into FAO, it was said, was a car with a CD number plate. Several general service girls married men who became directors. The wife of one of them who then rose to be director general was at one time a G4 secretary. Superficially, it may seem that the number of men in such an organization free to marry is small in relation to the number of interested women. Many men, however, though

81

officially tied, felt free in the FAO environment. Facing the chance of starting a new life with an attractive partner many have also freed themselves officially.

Working in an FAO unit that has continuing contacts with field advisers broadens the range of potential husbands. For a substantial proportion of field specialists a change in their living environment is one of the considerations that leads them to seek such work. Periods of isolation on field duty then sharpens their interest in domestic company. Helping such a man through administrative obstacles, typing his reports in a hurry and so on, can lead on to brightening his evenings in Rome and thence to an enduring liaison. Participation in a technical meeting at a field station shortens the odds much further. I lost one of my best secretaries at one such occasion. From churning out letters and documents she moved to high levels, entertaining as wife of a senior FAO representative.

Substantial out-of-office benefits can also accrue to a general service staff member without the commitments of marriage. There have been several for whom apartments have been bought by generous professional friends. At least one was sufficiently hard to change the lock after the property transfer was completed and let in another lover. Another general service secretary inherited an apartment in London from her ex-boss – in this case a woman – to whom she had been helpful. From her subsequent property acquisitions it would seem that she inherited a cash fortune as well.

MOVING DOWN

Not everyone can be successful or lucky. Inevitably there were some for whom FAO meant a negative phase in their lives. They did not like it and left. Others did not like it, but had not the courage to leave a stable post with pension to follow. One secretary in Commodities Division who wore expensive clothes in a provocative style would have been a princess in old Czechoslovakia. There was a woman who said FAO was a prison from which she did not escape until it was too late – for another career or for another husband was not made clear. There were some men and women who had nervous breakdowns. Love–hate relationships in FAO were common conversation. One man said FAO made him ill, and left with a pension. Years later he could be seen eating at restaurants nearby.

General service staff who left a post because of adverse relations with the supervisor or whose post was abolished became 'floaters'. They continued to be paid. They went on to a roster in Personnel and were put forward as candidates for suitable vacancies as they occurred. Meanwhile, they substituted for staff away ill or helped units with an exceptional workload. There was a general reluctance to accept 'floaters' for new continuing posts. Divisions responsible for lengthy field project reports had a pool of typists to cope. 'Floaters,' somebody said, 'should sit in the pool.'

Continued status as a 'floater' was the cue to leave with whatever pension entitlements had been accumulated. For the adventurous, assignment to a field project was an attractive way out. Two years in Afghanistan, for example, as a conspicuous newcomer in international circles could radically change one's view of life. A return later to FAO Rome with a new background and outlook would always be possible.

The ultimate way out was suicide and there were enough suicides in FAO to attract attention. Appearance of an official notice that 'Mr or Mrs ... had died suddenly' was the conventional indicator.

It was an interpreter who attracted the most attention. He hanged himself in an interpreters' booth of the main FAO conference hall. A visitor being shown round FAO commented on a peculiar smell, belying incidentally the popular view that FAO is always holding conferences.

The general service staff of FAO complained a lot. Work was unevenly distributed; others did not pull their weight; others were promoted unfairly; bosses were unworthy, insensitive, downright unpleasant; but few wanted to leave. In Buenos Aires, a very cosmopolitan city, I met an English woman who had worked thirty years in general service posts. 'I like it here,' she said, 'it reminds me of FAO.'

6

THE FAO WAY

HOW TO JOIN FAO

To join FAO one can apply to it directly, answer an advertisement, or be brought to its attention through a contact. The first approach is almost certainly the least rewarding. Letters of enquiry concerning opportunities to work with FAO go to its professional and general service recruitment offices. Their first reaction is to dispatch a pre-printed bland letter of acknowledgement. This may be the only one the enquirer receives. Details of the applicant should then be circulated to potentially interested units. It is a matter of luck whether the head of a unit is looking for a new staff member with the qualifications of the applicant. Of course, a secretary with mother tongue competence in two working languages will always attract interest. If there is no response, the application goes on to a computerized roster which can be drawn upon when vacancies occur. This will certainly be consulted from time to time; meanwhile, however, the applicant is likely to have given up hope and taken another job.

Application for a post set out in a specific advertisement has the great advantage that all applicants meeting these stated requirements must be reviewed by the post supervisor. References will be taken up on those short listed. If there is no candidate who is already known to be outstandingly suitable, several of the applicants may be brought to Rome for interview.

Recommendation by someone who is known and trusted by the relevant unit in FAO is the most successful route to joining it. The recommendation can be from another staff member, a university professor, or someone working in a parallel agency or government department. Critical for the potential chief in FAO is that the recommendation be from a source that is professionally reliable and whose judgment is respected. A very few FAO staff members are the offspring of FAO parents, but they are not employed there at the same time.

FAO administration lays great stress on the interview. 'Call them in for an interview if you have any doubts' is their dictum. I personally have doubts about the accuracy of interview judgment as against reliable evidence of previous performances. However, the World Bank's recruitment of Harold Evans stand to disprove my opinion. He was appointed head of its Agriculture Division in the 1960s when it had wide project responsibilities, yet was called for interview in error, he says; his name was mistaken for another candidate's.

In the first decades of FAO's advisory work, experienced candidates for field assignments came forward in large part from the British, Dutch, French, and Belgian colonial services, as they were replaced by nationals of countries attaining independence. They knew former colleagues who had gone on to FAO and could renew contacts with them. People coming directly from universities, however, faced the problem of how to obtain some relevant experience. This was eased by the establishment of volunteer programmes. The governments of Denmark and Germany set the lead in financing associate experts' posts whereby qualified young people could work alongside an FAO staff member, particularly on field assignments, without cost to FAO or the host country. After some years of such training they became excellent candidates for FAO posts. Many of the outstanding professionals in marketing came in through this channel.

The best course, then, for young men and women who want to work with FAO professionally is to seek experience in developing countries first. This can be via field studies undertaken under university supervision, or through voluntary work with a government or similar agency. In any event, do not give up. There may be less money around for aid programmes and in consequence fewer posts to be filled; even so, those responsible for filling these posts still maintain that they are scraping the bottom of the recruitment barrel.

As can be expected, recruitment procedures move fast for the jobs nobody wants. One woman arrived in Rome and applied to FAO for work on a Friday. She was told to start work the following Monday. She says she has regretted it even since – her boss was a slave-driver and a snob.

My own recruitment had its hiccups. It started off well enough; the Director of Economics said there was a place for me in his division. I wrote for an application form, as he suggested. My university said it would give me a good reference. I first received a letter saying that I would receive favourable consideration, then a letter of appointment subject to medical clearance. This was to be obtained from a local doctor who was to transmit the documents to FAO under confidence. The doctor told me I was fit, so I assumed I would be joining FAO shortly. My university took the same view and appointed someone else to my post. After several months without salary I cabled FAO to find out what had happened. My medical papers had been lost, I was told. 'I should take the examination again.'

Geographical representation

Appointment to an FAO post can be blocked by over-representation of the nationality concerned. Professional appointments in FAO are supposed to be in proportion to national contributions to its financing. The professional posts are assigned weights according to their importance. The largest contributor to FAO has been the USA long meeting one-quarter of its budget. Its nationals should therefore fill one-quarter of the posts. However, there were never enough US nationals wanting to work in FAO, after it moved to Rome, to fill more than half their quota. The Japanese quota was never filled. Some senior Japanese were found eventually for posts in Fisheries; this was an area of great Japanese activity.

So the representation mechanism has functioned primarily as a limit on the number of posts that could be taken by nationals of countries with an ample supply of qualified candidates for whom securing an FAO job was a coup. The UK and Netherland's staff quotas have often been full, even after the limit was relaxed in the interests of working efficiency to allow an additional 25 per cent. The Italian quota was usually full, but mostly through people in junior posts. Other countries tending to be over-represented were Chile, Egypt, and India, all with an ample supply of well-educated people, but a low quota. A modification introduced when many very small African, Caribbean, and Pacific countries came into FAO was to allow every country one staff member irrespective of its financial contribution. This would promote the feeling that FAO was their organization. It also extended a Director General's scope to use such posts as bait for votes at election time.

Geographical representation is sound enough in principle. In FAO it kept a rein on the inclination of unit heads to recruit too many people of their own nationality. It could also, however, force the appointment of people who were not the best available for a particular post.

Two men I recruited under the pressure of geographical representation were lively and interesting, a part-Chinese Filipino and a Lithuanian American. Both were good value and went on to higher things, but they were not interested in marketing. The American had strong links with the Vatican. One evening at his house there were not one, but two cardinals – one leaving before the other arrived. It was someone's good idea to appoint him Assistant FAO Representative in southern Sudan. The FAO man in Khartoum was Muslim; there should be a good Christian in the south, it was felt.

HOW TO LEAVE FAO

To get yourself dismissed from FAO because you are often away from your office, do very little work, and are generally incompetent is quite difficult. There are always people in personnel quick to understand that your boss is biased against you because of differences in national background, culture, or religion. After successive attempts to get you out have misfired, he will see the futility of the direct approach and go for a lateral transfer. You will be recommended widely as unsuited to his humdrum office, but with high potential for responsible work elsewhere. You may well end up with a promotion.

An estimable exit is by golden handshake. This can be two or three years' full salary just to leave a little early and wait elsewhere until your pension comes along. The best approach for this is to have your post declared redundant. In this way you side-step assessments of personal performance that might take a sour turn.

There was a man who threw away his job for his cat. It was a magnificent Persian. He had been accepted for a post in Mauritius, an island which had the same strict quarantine rules as Britain. A hurricane delayed the plane for twenty-four hours. When it arrived there was no one at the airport to clear the cat. There seemed to be no other course but to go to a hotel. Here, however, the cat attracted attention. The police came; perhaps they would have forgotten the matter for a hundred dollar bill. In this case the FAO man said it was not his fault – the Mauritians were not at the airport when the plane arrived. He had to leave the country.

Another man found his wife was very upset over his impending departure on a training assignment. He went to his director to ask if it could be postponed, otherwise she might have a nervous breakdown. The director had been brought up in the Muslim tradition. 'Then you must get another wife,' he said.

Taking on an FAO assignment was often, in fact, the cue for men to change their wives. One of the best advisers we had in marketing used a year's duty on a Caribbean island to leave one wife and take on another; we would not have got him, he said afterwards, if he had not wanted to end his previous ties. FAO normally stood well back from such personal matters. It faced a problem, however, with an Iraqi food technologist who had married an American when he was a student. On assignment in Indonesia where Muslim law holds good, he divorced her in the traditional manner – clapping his hands together outside his front door and announcing three times in a loud voice – 'I divorce you'. Should FAO administration recognize this divorce and list her successor as his official dependent? The American wife said no!

My own daughter married a Tunisian. The FAO staff lawyer teased me because I had not asked a bride price. For a fair-skinned girl speaking two languages it should have been at least 100 cows, 300 sheep, or 3,000 dinars. I was saved a lot of trouble by my inattention when later she left him; custom would require that I return the animals in full together with their progeny.

Workaholics and people who just like their job can normally expect to go on with FAO until their 62nd birthday – if they have been appointed to a continuing post. In the first instance they will receive an appointment for a fixed term: six months, one year, or two years. While the prospect of secure employment may be a factor in attracting and holding high-quality staff, FAO is constrained to limit its continuing obligations by the biennial nature of its financing. Indeed, the proportion of the budget going to support continuing staff has been brought down to less than 60 per cent because of the uncertainties of future financing. Staff with continuing posts can still be terminated if the director general is obliged to reduce his staff, but they would be entitled to compensation.

End of contract

At the top – division directors and heads of departments – everybody is on a fixed term appointment, typically three or five years. It can be extended or renewed, but it also runs out. Various directors and assistant director generals have had to look for new posts, or return to a former government or university position, when their FAO assignment was completed. This limits the damage of a mistaken choice. It also permits a new director general to bring in his own people to some top positions. The Director of the Nutrition Division when Saouma was elected DG was French. He may have expected his appointment to be renewed. His first intimation that this would not be the case was an automatic note from Administration on the formalities of his departure. Aghast, he rang up the DG to find out if there had been a mistake; he was told to leave his desk clear at the end of the month.

At one time, continuation of employment by FAO was the best of all possible worlds. Professionals with their 62nd year in sight applied for extensions. Some transferred to field assignments where they could continue to 65 if a government wanted them. In the 1980s the fashion shifted to early retirement. Many people had joined FAO during its phase of expansion in the 1960s. By the late 1980s they had been with it for twenty-five years. This seemed long enough; it also entitled them to a pension of around two-thirds of their net salary. Depending on the demand for their professional qualifications they could look forward to interesting work and substantial earnings as a consultant. For some, the dollar/lira exchange rate was a factor. If the rate prevailing during their last three years would lead to a high dollar pension, it was judged wise to take it and run.

The next step after retirement was to get back to one's desk as a consultant or temporary, earning a fee as well as the pension. The main constraint on this was that the total payment received should not exceed the salary paid previously. This came to an end around 1986. Thenceforward, FAO staff on pension could only be employed if there was nobody else who could do the job. This had to be cleared at the highest level.

ACQUIRING THE LANGUAGES

Announcements of vacant posts in FAO commonly carry the wording, 'Applicants must be fluent in one of the working languages of FAO (English, French, and Spanish), and have a good knowledge of at least one other.' This could lead a monolingual candidate to give up hope. Certainly a professional in the technical subjects of FAO will find a command of French and Spanish very useful. These are languages used by a large number of countries that FAO helps. Without a working knowledge, a technician will be barred from important areas of professional interest, or disadvantaged by dependence on translation. For English mother tongue people who are accustomed to having others learn their language, and for the many Africans and Asians who use English, acquiring other FAO languages can be a formidable task. Help in interpretation and translation is available. Still, it is convenient to be able to skim material in French and Spanish for the main points, and to check whether translations from English have kept the sense of the original. It is also useful to be able to maintain a fair conversation with people who speak those languages better than English. Marrying someone speaking a language one needs is a very positive approach. Extended stays in countries using French and Spanish can revive an earlier acquaintance based on formal instruction.

To succeed in FAO, however, one must acquire facility in its own super-cautious circumlocutory jargon – 'FAO-ese'. I knew I was there when my relatives said, 'Stop talking to us, John, as if we were a bunch of foreigners; don't be so pompous.' Slang and colloquialisms had vanished from my vocabulary; they might not be understood. In had come guarded, elaborate expressions, clearly presented. Even so, my writings were still too direct for some FAO uses. The head of the Economic and Social Department got where he was, said a one-time colleague, because when he cleared a paper no one country delegate would be able to see an adverse reference. The following excerpts from K.A. Bennett's rewrite of *Little Red Riding Hood* are in FAO-ese of the 1970s. With their evident wildlife concern they should still rate with the 'Greens' of today.

> Little Red Riding Hood decided to initiate an undertaking whereby she could establish visual and aural contact with the mother of her mother who was less than completely well. She decided, moreover, when visiting with the mother of her mother, to diverge from the usual output–input pattern that governed disposal of surplus small-cultivator produce surplus, that is, to the immediate needs of Red Riding Hood's family. Instead, she would offer some eggs to the mother of her mother without an expectation of economic benefit.... There in bed was the mother of her mother (or so Red Riding Hood thought) wearing her bed-jacket and sleeping bonnet.
>
> In fact, however, it was the Big, Bad Wolf. For some time his diet had been suffering from a certain rigidity and imbalance, and only the previous day, feeling particularly deficient in all the proteins and most of the vitamins, he did much to re-establish a well-rounded diet for himself by entering the cottage unobserved and gobbling up the old lady.

Oxford and Webster

One must also know which English is official in FAO. It began in Washington. Its first director general was British, but FAO used the American vernacular. It continued to do so when it transferred to Rome. The first two directors general there were American. In Rome it became dependent on secretarial staff who were English and Irish in educational background. Only a few of its American general service staff came over. A considerable effort was needed to maintain a consistent American spelling and usage. Officially, the organization moved to 'English English' with Dr Sen. He was accustomed to it in India. He also saw that a large proportion of the newly independent countries joining FAO would use it too. This gave a freer rein to the more literary English styles. One professional colleague took special pride in his letters. He couched a letter of termination to one of our field advisers, with due appreciation of work done, in such elegant prose that the man, who was using English as a second language, thought he was being complimented and asked for a promotion.

A very literary Englishman, Henderson was head of the Publications branch. At his retirement party his colleagues presented him with the usual farewell gift. For a man with his interests they felt that it should be a book. They were giving him, they said, the one book that they were sure he never had – *Webster's Dictionary*. For two decades the *New Oxford Dictionary* was for FAO the authority on English usage. Now Webster is in favour again, even in some English academic circles.

French took a new life in FAO with the election as DG of Saouma. He came from Lebanon where French was the language of educated people. As a counter to the prevailing use of English in international discussion he was a rallying-point for the French. In the 1987 election they voted for Saouma against Moise Mensah of francophone Benin. Mensah spoke English too well. While Saouma speaks English, he prefers French. One of his first steps on election was to set up a post titled 'Chef de Cabinet'. French and also Spanish could be used in his weekly meeting with senior staff. Those who spoke French fluently were thought to be nearer to his heart; their own view was that finally they were relieved of the handicap of having to use English. A francophone chief of Library said to me at such a meeting: 'Now, finally, I can express my thoughts before the moment has passed.' This facility can become very expensive. A team of interpreters had to be provided for this weekly internal meeting – simultaneous interpretation English – French – Spanish at a full-scale conference requires twelve interpreters, four for each language allowing for one in each direction, and rotation to keep the individual workload to four hours per day. Interpreters' salaries in 1988 were $200 per day, plus travel expenses and per diem. Where draft resolutions or reports had to be prepared and presented during a conference, the services of translators would also be needed. The interpretation and translation support cost of a trilingual meeting could be $4,000–5,000 per day.

Adding Arabic and Chinese has augmented these costs. Arabic became an official language of FAO following the election to DG of A.H. Boerma with strong Arab support. Chinese was an original language of the United Nations. It became official for FAO when China rejoined in 1973. It was to be used in official meetings where China participated.

At the technical level, interpretation costs have forced units to hold seminars, etc., separately for each language group. This allows more scope for nuances in the exchange of experience. However, it sacrifices one of the roles of the international agency – that of bridging communication gaps between developing countries. It leaves with the organizers the responsibility of transmitting to the participants information on relevant experience in other language areas.

Efforts are made to include in each technical support unit staff with the languages required. This is not easy. Candidates fluent in French or Spanish and also qualified to give a lead in marketing development were especially scarce. It was not a subject much studied in Latin countries. After long searching, a really good young Frenchman was found for rural finance. He was drawn off to the World Bank within a year. In the more traditional areas of agriculture and forestry this may have been less of a problem.

Language-training facilities were available at FAO. They were used intensively by secretaries who received an extra allowance if they could pass a test in a language other than their own. These courses were the first casualty when funds ran short in the 1980s, although some of the instructors continued them on a private basis. Equipping a technical specialist to work in a language that he or she did not know was more difficult. There were crash courses – total immersion for a month – but the Frenchman we sent to Tanzania after such a course had to come back; his English was still inadequate.

FAO spends a lot on its documents because of its obligation to issue them simultaneously in various languages. Governing Conference and Council papers, and priced publications, must go out in English, French, and Spanish. The bulk of these are prepared in English and translated.

To the author it was often galling that the English version should be held back a year or more until it was ready in French and Spanish. One valuable lesson was to leave out of a text any references that would date it. Some FAO monographs have been reprinted without change for over twenty years.

Writing in a clear, simple style is especially important where texts are to be translated into other languages. Issuance of printed texts in Arabic and Chinese is undertaken at the request of the countries concerned. FAO books are translated and published in many other languages with the organization's permission.

A continuing brake on the publication of technical material by FAO has been the practice of charging to the sponsoring divisions the staff and other overhead costs of the publication unit. Thus in the early 1980s the cost to a divisional budget for publishing a manual of 200 printed pages in three languages was around $100,000. Usually such an amount could be raised only by spreading it over several years or drawing in savings from other units that did not have their material ready in time. Perhaps it was a healthy constraint. Technical manuscripts were often prepared by consultants concerned to satisfy their peers. It was the task of a supervisor experienced in Third World readers' requirements to pare this material down to basic essentials. A further consideration is that FAO should not duplicate material available from commercial publishers, many of which now specialize in works for the developing countries.

FAO's list of books in print has over two hundred advisory texts and situation/statistical report series. They are well adapted to developing country needs, and prices are low to moderate. Getting them to the people they would benefit is hard, however. A quota is supplied free to all member governments. Mostly these go to a designated department in their ministry of agriculture. Often they stay there.

People not close to this department and in local universities and training institutions, depend in practice on direct contacts with FAO personnel and programmes. In Rome FAO has its own bookshop, well placed near the main entrance. Its list of publications is available at its country representative offices, but books are not on sale there. FAO book sales agencies have been appointed in eighty countries. They can often prove to be off-hand or obscure. On a recent visit to Her Majesty's Stationery Office, the FAO agent for the UK, I noticed that they had no FAO books on display. Lack of advertising by the FAO, and thus low demand, was the reason given. The FAO bookseller in Cairo some years ago had the bulk of its stock in piles on the floor. Potential customers moved around bent sideways from the hip to read the titles. Even so, many FAO books have sold in thousands and have been reprinted time and again to meet the demand.

START A WORKING GROUP AND WRITE UP PROJECTS

Having had an opportunity to see colleagues in action and to appreciate the way FAO works, an aspiring professional will see the advantages of promoting his or her own working group. It should meet annually, or at least biennially, preferably calling for participation by governments in various parts of the world. This will bring the professional continuity of employment and justification for travel to far-away capitals. Tropical commodities are well adapted for this – they are grown from Ecuador to Papua New Guinea. Agricultural credit is another good base; its problems are never-ending and they can be treated regionally, sub-regionally by language groups, or globally with interpreters.

When the group meets, various people can be briefed to bring up topics on which they need further studies. With their endorsement in hand the FAO manager can apply to Administration for 'further resources'. With successive expansion will come a promotion. With such a group behind you, you are safe from the carping of jealous colleagues. You are not building an empire for yourself, but responding to external requests. The governments will back you up too; for they have officers interested in periodic visits to Rome, Bangkok, or Nairobi. Institutional issues such as the development of agricultural co-operatives or credit services are also a good basis for regular intergovernment meetings: they are more easily resisted, however, by the budget controller, precisely because they are more continuous. Their urgency is less easily evidenced in contrast to newspaper headlines about low commodity prices or, alternatively, extremely high prices that will discourage consumption and promote the use of substitutes. Institutional growth has the advantage over commodity issues in any appeal to external donors. The problems of hard fibre exporters will seem very distant to non-governmental donor organizations in Scandanavia. Agricultural co-operation and credit will always attract their support.

Regional conferences are a classic vehicle for promoting a new continuing activity. Few headquarters professionals get a ticket to attend a regional conference without a divisional mandate to secure a conference resolution that will bring work for him and his colleagues. The in-fighting starts when the ADG in charge of the conference sets a cap on the number of resolutions with budgetary implication. This, indeed, sharpens the competition. It is not enough that the conference urges the FAO to give early attention to a specified issue as funds permit. There will be pages of such bland evocations. A minister or influential government representative must be induced to catch the ear of the ADG concerned. If a short field trip could be arranged to fix his attention on the matter, so much the better.

A good working group has very enduring qualities – there are some that have spanned the decades and still continue. A working group can also be inherited. New appointees can find themselves 'servicing' such a group without much effort. Since the government participants will also have changed, you can short-cut your preparations by adapting papers presented at earlier meetings. This will not be noticed; the problems remain the same – and solving them is in nobody's interest. The defence for all this is that governments welcome a forum for discussion; they look to FAO to provide one. If it does not do so they look elsewhere. The establishment of UNCTAD and a half-dozen single-commodity organizations stands witness to this.

An ongoing expectation of technical support officers is that they will bring in new field projects. This is done by helping governments formulate their requirements for technical assistance. FAO is not blatant about this. In the World Bank, promotion at one time depended almost entirely on how many loan projects someone could prepare, and how large they were. A colleague was on a mission to Thailand to prepare a seed development loan. The team carefully estimated the investment needed, and their leader telephoned his supervisor in Washington to report progress. Instructions came back – 'Too small; double every figure.'

In FAO at one time the game between technical divisions was to orient a field project so that a particular discipline was predominant. This carried substantial kudos. It determined which division should operate the project; it also brought in additional funds via the overhead allowance. Part of this went to central administration, part went to the operating division where it could be used to pay for additional professional and secretarial support. With the establishment of the UN Special Fund for Development this became a strategic FAO activity – as satirized in John Spears' inspired lyric:

'Twas on a Monday morning that the DG said to me,
"Will you write a letter to the Fund telling them that we
Are preparing our new project for a Member Government?"
So I took a piece of paper out and this is how it went:

Estimate the project cost and see what I can do
If I add a million dollars then I multiply by two
Count up all the experts multiply by three
The more the experts in the field the bigger job for me.

'Twas on a Tuesday morning – imagine my elation
When they asked me if the project needed any irrigation
So I flood the project area quickly providing that they oughter
Make the operating agency entirely Land and Water.

'Twas on a Wednesday morning that a letter came to me
Asking if the project needed any forestry
So I quickly got to work and using my imagination
Made the object of the project one of reafforestation.

'Twas on a Thursday morning that I nearly did a dance
When they asked me if the project was of interest to Plants
So leaping on the project like a bureaucratic vulture
I quickly turned the project into one of horticulture.

'Twas on a Friday morning that they asked for my decision
Was the project of significance to Animal Division
So ignoring everybody else I took another sheet
And wrote a brand new project with the emphasis on meat.

It was logical that he should go on to the World Bank. In 1987 he was head of its unit on 'environment', the currently fashionable subject.

The risk of projects being distorted by technical officers seeking operating status for their divisions was indeed there. The projects might then be neglected by staff more interested in their professional field than in project management. This led to the establishment in FAO of specialized project preparation and operation divisions.

TAP THE ALLOWANCES

FAO has followed the practice of diplomatic representation in according its staff privileges in compensation for their being away from home and to recognize their international status. These include conditions of duty travel, home leave every second year, educational and family visit travel, children's education allowances, and exemption from car import, registration, and fuel taxes. Attitudes toward these privileges among FAO staff have ranged from careful exploitation of every opportunity to unconcern inspiring friendly reminders of their impending lapse.

In the 1950s FAO professional staff travelled first class on a par with personnel of comparable rank in government service. On home leave they could opt to travel one way by boat. As their numbers increased and air travel became more intensive, such standards were recognized as unrealistic. First class air travel was restricted to the highest ranks. Professionals travelled business class on intercontinental journeys, otherwise economy class.

As the destination for home leave one could nominate the place of recruitment, one's home country, or a third country should there be political reasons why the home country was not convenient. A colleague recruited from the University of Hawaii felt that the opportunity for free travel back to Hawaii was too good to let go. To reach Honolulu one could go one way and complete a round-the-world-trip on the return. It became tedious, however – to qualify for home leave he was obliged to spend ten days at the official destination. In his case this had to be spent in a hotel; his friends and relatives were elsewhere.

In the years alternate to those in which one is eligible for home leave there can be a paid family visit. Children attending school in the home country can come out, at the organization's expense, to be with their parent, or the parent can travel home to spend annual leave with children.

People who are really interested in their work can use home leave travel to visit projects, contact governments, or attend professional meetings *en route*. They can also take opportunities to contact people who have done relevant research, to visit university professors who could recommend bright young people for their unit, to lunch with consultants, and others with useful experience to exchange. Such a person may also convince their family to spend part of the leave with them at some small university town where there is a conference in their subject area. For others, leave of less than a month is useless – they need that long to get FAO out of their soul.

Professionals in FAO can import a car free of duty every four years. At director level they are entitled to CD plates and exempted from the Italian road tax. This privilege is cherished, especially, it is said, by wives. It may also relieve them of fines for minor traffic offences. Tempting, also, is the allowance on leaving FAO for the transport home of personal effects. Many people have acquired new items for this – to make full use of his weight allowance one man included cut firewood.

SERVICES THAT MATTER

Ensuring that the staff at FAO headquarters drawn from many nations and cultures had a satisfying midday meal was vital for FAO. It was also an essential service to visitors. From the beginning there was a self-service cafeteria and a restaurant with waiter service on the eighth floor of Building B. When Building C was added they were supplemented by 'the Grill', offering a limited range of hot dishes cooked to order. Snack meals have also been available from two other cafés, and from one run by the staff co-operative. There were staff representatives on the Committee, to which cafeteria menus and prices were submitted for approval.

Hundreds of FAO staff used these services every day. Their comments alternated between complaints of poor food (especially from the French) and of high prices. All agreed that the cafeteria was a good place to meet one's friends, also to sight visitors of whose arrival one might not be aware. Seating was informal at long tables; groups could form without pre-arrangement. For those with religious and dietary food positions the cafeteria was a godsend. They could see – approximately – what they were ordering without having to cope with Italian gastronomic terminology.

The restaurant began on a near-luxury level with a superb trolley of hors-d'œuvres and Greek lemon soup. It had a clientele of senior staff talking to each other, middle-level staff hosting colleagues from the field and out-of-town visitors, and less senior staff celebrating a birthday or a promotion. It was very good. To fill a gap in conversation there was the view of St Peter's. Sadly the restaurant went downhill. FAO catering was opened periodically to competitive bidding, and new enterprises offered more favourable terms, then reduced the quality. There were strikes. In 1989 the once-glorious FAO restaurant was featuring a 'special menu' for L10,000 (about $7.00), less than many would pay for self-service.

The FAO commissary was a titillation. For FAO staff with professional salaries it was not important. It was a convenience where they could buy some things they wanted without having to look for them in the shops. If there had been no commissary, local merchants would have set up to meet the demand. The price advantages were significant on tobacco, drinks, and butter – all things, someone said, 'that one should forego'. Even the drinks were available in local shops at prices much lower than in the UK, for example, where many of them originated. American breakfast foods and English biscuits could also be found locally, but more expensively.

For those who were not allowed into the commissary it was a palace of wonders. FAO staff were forever pressed by Italian friends and acquaintances to 'get us something from your commissary; we will pay for it'. Those who were allowed only a limited range of items envied those who could buy more. The Italian staff of FAO were not allowed to use the commissary initially; they complained of discrimination until they got access.

Embassies and international organizations are often allowed to import items normally consumed by their staff that are not easily available in the host country. Initially the FAO commissary was small; orders had to be presented in writing, and purchases were picked up later in a discreet brown paper parcel. Some time in the 1960s the FAO commissary became a supermarket. For FAO management, commissary prices became an element in wage bargaining over the cost of living. For those staff who maintained their ideals the lines of trolleys loaded with food packages were offensive. At one stage each FAO cleaning lady bought a bottle of Chivas Regal whisky monthly. She then made a dollar or two reselling it to a man waiting outside. The volume of sales eventually became an issue with the Italian government. Italian butchers, it was said induced frontier guards to hold back consignments of Danish meat destined for the commissary. Finally, the range of goods offered was brought back to those not easily available in Italy, and the hours of opening were restricted.

FAO has a credit union for staff with money to save or wanting a loan. It grew out of the cash needs of staff newly arriving. Previously they went to FAO Administration for salary advances. The Credit Union began as a staff co-operative paying rent to FAO for the use of a room. Later, to avoid possible Italian tax obligations, it became an integral part of FAO. Loans are related to expected FAO earnings; repayments together with interest should not exceed one-third of salary over the expected duration of employment. Thus the Credit Union is unlikely to provide the money to buy an apartment; it can only top up funds available from other sources.

Shirley Phillips was a founder member of the FAO Credit Union and its Secretary for the first six years. She also started the FAO staff co-operative - originally to charter low-cost flights for the holiday seasons. It grew into the present Staff Cultural and Social Association with 'interest groups' ranging through art, music, theatre, dancing, meditation and various sports, to work opportunities for spouses. It helps new staff find accommodation, retains an English-speaking lawyer for free consultation and runs its own café and lounge. It has talked of a club where FAO staff could meet out of working hours. Together, however, FAO and Rome offer so much that after the first few weeks most newcomers are absorbed in a new life.

Much less evanescent is the security clamp on the FAO doors. For many happy years they were open to visitors and friends; only the evidently ill-intentioned were excluded. This ended with the hijacks of the 1970s. Thenceforward entry has been conditional on presentation of a special card; visitors must deposit a passport or similar vital document and proceed with their status labelled on the breast.

In the 1950s Major Woodcock, retired from the British army in Italy, sat alone guarding the FAO entrance. He knew everybody. To help those needing accommodation he set up, with a Sicilian countess, the Woodcock Gravina Pensione. It became a home from home for people in from field projects. They sat together at long tables for dinner; wine and a liqueur were on the house. There, newcomers had their best briefing. The pensione was continued by two FAO wives as the Lancelot. The man who financed them said his profits made an FAO salary seem peanuts.

7

FAO IN ROME

'Is it true that FAO is leaving Rome?' asked my *portière*, wondering on somebody's behalf whether my apartment would become free. 'Are we really going to Munich?' I was asked repeatedly by secretarial staff with local boyfriends around the time its Olympic buildings were on offer. 'They won't really move to Delhi will they?' was the hope voiced when some member of the FAO Council said this could be the answer to rising local salaries in Rome. The knowing and the cynical saw these rumours as Administration leaks designed to dampen down the staff union in its pursuit of higher pay or to nudge the Italian Government into providing more space. Its original offer was a vacant office block in the centre of Rome, built to house the administration of Mussolini's African empire. It remains the property of the government of Italy, rented to FAO for one dollar per year. An additional building to house large meetings and the Lubin Library was constructed in time for the next Conference of FAO member countries. Further space was made available later by more building and by vacation of offices occupied by the Italian Ministry of Post and Telegraphs. The grand marble staircase of Mussolini's architect finally came into its own. In 1990 there were FAO Buildings A to E on the main site. F Building, housing the Fisheries and Forestry Departments, was rented space on the Viale Cristoforo Colombo, linked to the others by a frequent shuttle bus.

At one of his weekly staff meetings DG Saouma drew a verbal picture of a grand new agri-complex to be located on the way to Fiumicino airport. It jarred on his senior staff who were attached to the Circo Massimo site with its central position and unique historical surroundings. Perhaps FAO's fellow UN agencies in Rome were also cool toward the idea; shortly afterwards he announced that a new building replacing the adjacent school to the east would be sufficient. In late 1989 plans were announced for a new structure on the garage site, the replacement of

Building E and additions to other buildings, in order to accommodate 700 more people. Building F would then no longer be needed. Within a month, work came to halt with the discovery of an early Roman villa.

CHOICE OF ROME

The original founders of FAO felt that it should be located at the same place as the United Nations. In fact it started in Washington because the Interim Commission had established a base there, as decided at the Hot Springs Conference. This was temporary, however. When the United Nations settled in New York on land given by the Rockefellers it became evident that a different site was needed for an organization concerned with agriculture, forestry, and fishing. This came up for a vote at the fifth FAO Conference in Copenhagen. Five ballots were needed to obtain a clear majority for one place. Sen, on the Indian delegation, favoured Rome because of its central location and convenience for air travel. The opposition wanted to stay in the USA. Denmark had offered its castle at Elsinore, but settled the deadlock by voting for Rome. The Government of Italy offered an attractive building on very favourable terms – a rent of $1 per year. That FAO's predecessor organization, the International Institute of Agriculture, had been in Rome must also have been a point in its favour.

Professor Ugo Papi, Secretary of the International Institute, was on the Italian delegation. He claimed that his campaigning tipped the balance – and he could be very convincing. At one FAO conference, he spoke on that year's report on the state of food and agriculture. For the first eight minutes it was the best speech of its kind that I had heard; on this occasion he went on for a further twenty minutes and destroyed the effect.

The returns to Italy on its offer to FAO have been immense. For every dollar paid in annual contributions some sixteen came back in staff salaries and contracts, leaving aside other local spending. In one year the Italian Government's share of FAO's regular programme budget was $5.7 million. Italian staff received $73 million in salaries. Italian firms obtained $10 million worth of orders. In comparison, the cost benefit ratio for Belgium and the UK was 1:7, for the Netherlands 1:5, and for France 1:2 (Hancock 1989).

FAO moved physically from Washington to Rome in 1951 on the ships SS Vulcania and Saturnia. Two-thirds of its staff decided not to make the transfer, leaving many posts to be filled. The first crossing on the Vulcania was rough indeed. Along with other upsets, the deputy director general lost his teeth being sick over the side of the boat. Crawling to his bathroom, he ended up with the toilet seat around his neck.

How far has FAO benefited from being in Rome and how serious have been the disadvantages? This is a recurring theme of debate between those who stayed with FAO and loved it, and those who left in greater or lesser part because they 'never settled down' in the local environment.

THE CHARM OF CARACALLA

The Emperor Caracalla had an ugly reputation; the part of Rome which bears his name is striking in its beauty. FAO is blessed with green on three sides. In front is the avenue of great pines leading to the Arch of Constantine, the San Gregorio church with its soft muted colours, and great banks of trees. To the south, the Viale delle Terme di Caracalla and the ruins of the great imperial baths maintain the same aura of grandeur. There are also private gardens. One has a sheltered lawn; another has a large kaki (persimmon) tree which looks dramatic in autumn with golden fruit hanging on bare branches, a classic portrayal of the land of Cockaigne. The Circo Massimo to the north, famous for its chariot races, remains a great open space still predominantly green, backed by the red-brown walls of the Augustan palaces. FAO's site is also very convenient. Great roads converge on FAO from four directions. The metro station hidden below offers fast and easy access to a major residential area; the bus provides direct access to others.

At the entrance to FAO stands a granite pillar. Doors are framed on each side with nine levels of windows carved in the style of the rock carvings at ancient sites in Ethiopia. The pillar was brought from Axum by Mussolini in the 1930s, emulating the Roman emperors of the first century. At one point the Ethiopians wanted it back. Another view is that it better advertises their country in its present position.

FAO's main buildings include a number of rooms decorated in the style of particular countries and bearing their name. Features of these rooms range from murals depicting scenes of national agriculture to carved woodwork and symbols of religion. The Thai room is guarded against the entry of evil spirits by a fine set of temple lions at the door. In the Indian room I have endured tedious meetings whilst contrasting my lot with that of marching peasants on the wall facing me. In some rooms, national character was more skimpy; one had to look around carefully to see what justified the room's name.

The most elegant of these rooms was that of Morocco on the eighth floor of Building B with direct access by special lift. It featured a small white marble fountain and was ideal for receptions in summer. It is favoured by the Director General for his more select parties and influential guests.

If for no other reason, FAO's location in Rome is justified by the view from the terrace on the roof of Building B. So many people have been able to enjoy it. There are the Alban hills to the south, green in the main with white patches marking the Castelli towns. To the east, in late winter, one can see the snow glistening on Mount Terminillo, promising good skiing within a two-hour drive. The dome of St Peter's and those of a score of lesser churches stand in clear view to the north. In autumn the swirling of great flocks of starlings over the Roman skyline brings a lump to one's throat. The terrace, one proclaims to visitors, is FAO's greatest asset. From the restaurant in the UN building, New York, one looks out on to the East River, bright and clean on a winter morning. The Palais des Nations, Geneva, has its own green park with mature trees and peacocks strutting across the lawns. No other UN agency, however, has a view to match that from the terrace of FAO.

The terrace is also a great amenity for FAO staff. After a quick lunch at the self-service cafeteria one could then stroll up and down the terrace for a while. It was exercise, a breath of fresh air, and an opportunity to talk to friends. One also met people there with whom one had business; it could be settled with a few words much more easily and pleasantly than telephoning through secretaries or going to the other person's office. Visitors from the countries FAO was assisting were also on show there and advantage could be taken of this to contact them. Without the terrace one might not have known they were in Rome. Lunch-time on the terrace is often quite productive.

There are some negative aspects, however. The common reaction of visitors on seeing several hundred FAO staff members parading up and down, or with their legs stretched out from chairs on a sunny day was apt to be that they should all be in their offices, hard at work. This was certainly the impression made upon a visiting group of United States senators on one occasion. The following day, instructions came from the Director General's office that the chairs were to be removed. People would not give such a relaxed impression or stay on the terrace so long if they had to stand all the time.

Like the Golden Gate Bridge, Mount Fuji, and some other high places in the world, the FAO terrace has also tempted the suicidal. Twice in recent years people have leapt to their deaths from its parapet.

One was a secretary in personal despair, the second a refugee from Ethiopia invited up there by friends.

FAO's great neighbour in Rome, the Vatican, has certainly added to its stature, if not to the success of its endeavours. For the large Roman Catholic populations of many countries, Rome is the centre of the world. For the United Nations agency responsible for food and agriculture to be there too adds to its aura. This is recognized in FAO. One afternoon of each biennial governing conference is assigned to an audience with the Pope. Periodically he is invited to grace the conference and to address it. That he would not ease the task of feeding growing populations by using his influence for birth control has helped keep the Director General in business, a cynic might say. The comment of Earl Butz, the US Secretary for Agriculture, whose salty jokes may finally have lost him his job, was, 'He can't set the rules if he is not in the game'.

Rome is also very conveniently situated in the world. FAO benefits from this professionally. There are direct flights to most places in the Third World – in contrast certainly to New York, Washington, and Vienna. Travellers from North America and northern Europe heading for Africa and Asia can stop over conveniently in Rome. This helps collaboration with other international and bilateral aid agencies and the sharing of information and experience.

On facilities for living, Rome is outstanding. For decades, single people who would have had to share apartments in London could find a spacious one for themselves in Rome with the luxury of a terrace and one's own flower pots. Rent fixing and inflation have since driven a wedge between the lucky in possession and the unfortunate newcomer. Still, housing around Rome is much cheaper than in Geneva, London, or New York, and the unlucky newcomer may qualify for a rent subsidy. Schools using a range of languages are available for FAO staff children. Working mothers can take toddlers in the car to work, hand them over to kindergarten supervisors at the doors of FAO and pick them up there again in the evening.

Maids are a more difficult problem. Two have lasted me over a thirty-year stretch. Faustina, tall and blonde, came from Tuscany where my landlady the Marchese Misciatelli had property. The second, Leda, was from the Rietana, north of Rome. She kept my parquet polished to danger point and the terrace flowers well watered. A minor penalty was the nigh-immediate disappearance of any sweet liqueur. In contrast, I know one FAO working couple with two small children who are on their thirty-first maid. In the late 1980s Singhalese maids were in fashion. One couple came from Sri Lanka where the husband worked in a bank. He took a year's leave to work in Italy. Together they earned enough as cleaners to build a house when they returned home at the end of their stay.

Driving to the sea for a swim in the evening is out now, likewise driving anywhere that involves returning to Rome on Sunday evening. Far too many cars compete for the same square metres of road space. Ostia has become a residential suburb with a polluted sea. Romans still bake on the beaches, however, spraying their bodies with water which they bring with them.

A house in the country is now fashionable, ideally in an old hill town or by the sea some distance away. The drive out is on Friday afternoon, directly from the office, the return on Sunday afternoon or after midnight. Beyond commuting distance such houses are fairly easy to find. Parents acquire them as a place to take young children.

Sports facilities in Rome can also come with unique associations. The golf club along the Appia has the Claudian aqueduct in the background, soft brown and red against the green, with the Alban hills rising in the distance. Less expensive courses can be reached in an hour. The entrance to the tennis club at the beginning of the Appia Antica is where St Peter turned back to be crucified. We played cricket in the Villa Pamphili amid majestic pines and baroque statues – the English-born Prince Doria Pamphili was honorary chairman of the Commonwealth Cricket Club of Rome. Matches there were featured on British television. Sadly they came to an end with the transfer of ownership of the villa to the city of Rome and its conversion into a public park. One can sail a small boat on Lake Bracciano, keep a sea-going yacht at Anzio or Fiumicino. More up-market are Porto Ercole and Punta Arena along the Tuscan coast. The scope for visiting Etruscan, Greek and Roman antiquities is unending, the best of them all within a day's travel.

Delegates from member countries have always liked coming to FAO in Rome. It is a city geared over the ages to receiving visitors and making them welcome. To say it offers good opportunities for shopping is a classic understatement. For the evening stroll there are the cafés of the Via Veneto. Those with a taste for the macabre can see in the same street skeletons of Capuchin monks arranged imaginatively by their successors.

FAO has also been hospitable. There is always a welcoming cocktail party to open a meeting, perhaps others offered by embassies with representation funds. Regular visitors develop personal contacts with FAO staff. One Lebanese delegate made a practice of renting a house at Ostia for the duration of the FAO governing conference so that he could entertain there undisturbed.

On dress, FAO has always been informal. Without coat and tie one can work fairly comfortably through the Roman summer. In Washington, where it can also be hot, Ralph Phillips felt a rebel in an open-necked shirt; colleagues from Civil Service backgrounds stuck to black coat and striped trousers. A British Chief of Information in FAO continued this practice to the point that one wondered if it was for protection – as appeared to be the case from the manner of his departure. Deputy DG Sir Herbert Broadley was accustomed to wear a dinner jacket for some evening occasions. In Rome he had to give it up, he was so often taken for a waiter.

SOME DISADVANTAGES

A continuing feature of life and work in Rome is the high risk of theft. One learns to live with it and discount it. As one who has had his pocket picked three times, car stolen five times, and places of living broken into at least ten times, I speak from experience. The saving grace is the low risk of violence. Physical encounter between property owner and thief is very rare – thieves in Italy are too intelligent for that. When my pocket was picked at a railway station I never knew about it, reaching a hotel in Capri before I found that the bulk of my money had gone. I was awakened one Sunday morning by the crunching sound of someone breaking in my door. The intruder expected me to be away for the weekend; when I shouted, he ran. Thieves can also be very silent. One FAO couple were robbed of a fur coat and jewellery on a Sunday afternoon while they were watching television in another room. Women can be particularly inconvenienced by the prevalence of handbag snatchers. These thieves count on surprise, coming up from behind, snatching the bag, then leaping on to the back of a motor cycle before the owner realizes what has happened. Not everybody has the presence of mind and determination of my first secretary who, seeing the man with her bag jump on to a tram getting into motion, stood in front of it shouting '*ladro, ladro*' (thief, thief) until he dropped her bag and ran.

Three American women home economists approached by thieves resisted in unison. It ended with one of the male attackers shooting himself in the foot. Usually the documents and keys in the handbag constitute more of a loss than any money. Tedious journeys and procedures are required to replace them, locks have to be changed, and so on. A woman holding on to her handbag can be dragged to the ground and hurt: however, intentional violence is rare. The succession of murders attributed to the Mafia and the N'dranghetta in Calabria which give Italy a very bad press have not involved FAO personnel so far, nor have they or their family members been thought worth kidnapping for a ransom. A van bringing cash to the bank in Building B was once held up in the FAO courtyard. The gang pointed their 'mitras' at people nearby, but they did not fire. Still, it is worth observing the old Sicilian custom of never sitting with one's back to the door in a restaurant, just in case an executioner comes in and make a mistake in identifying his intended victim.

A saleswoman for large Swedish cars told me that many of the American and Canadian embassy people she contacted said they would not buy a car at all while in Rome; the traffic and the parking were too much of a hassle. Years ago Director General Sen had announced that he would not attend early evening receptions at the time of the biannual governing conference – it meant he spent too long in the traffic.

Determining where a car will be kept is an essential step; few buildings in the centre of Rome have their own garage. To get to where she keeps her Jaguar, a neighbour of mine takes a $5 taxi ride. Many evenings I have worked on in FAO or sat with friends in a bar until after 7pm when parking the car near my home becomes practicable. Where to put the car at FAO is also an issue. In the early days parking under one of the bridges between Building A and B was the goal. There, an open car was shielded from rain and sun. Later, getting into the FAO parking area at all was an achievement. Here the risk of theft was less. A commonly-heard comment regarding cars left outside the parking area was, 'I'll give you a ride if my car is still there.' Sometimes there were self-appointed guards. It was never clear whether these were agents for the thieves or just small-scale extortioners sticking long needles into the tyres of cars whose owners would not pay up. The competition for parking space worked strongly in the organization's favour. It meant that several hundred staff members reached the building well before the official hour for work. One woman felt so strongly about her parking place that she rose at 6.45am every morning over a twenty-five year period to be sure of it.

An advantage for people driving into FAO from outside Rome as against many other cities is that the rush hour is diluted. State employees arrive for 8am, and shops open at 9am. State employees go home at 2pm, bank and FAO people at 5.30pm, and the shops close at 7.30pm.

According to some criteria the atmosphere in the centre of Rome is now polluted. Street cleaners, traffic police, and young people on motor cycles wear white masks over nose and mouth. Air pollution is worse where streets are narrow and the buildings high on each side. FAO is not yet affected.

Lack of air conditioning in some of the main FAO buildings makes for uncomfortable working conditions in July and August. DG Sen once defended this in public by saying that in his country people had to endure a much longer period of heat discomfort. Access to air conditioning was very limited even in Saudi Arabia until the 1970s. I was once guest at a formal dinner in the royal chicken-house at Damman – it was the only air-conditioned public building in the area. The origins of this air-conditioned oasis are curious. The king of Saudi Arabia had once enjoyed a chicken meal offered by an American petroleum company, and expressed the wish that he could have his own supply. Eager to please, the company assured him that it would be done. Poultry specialists were brought in from California, who declared that the heat stress in the Saudi Arabian summer would be too severe; there must be air conditioning. So the company built an air-conditioned poultry house and stocked it with prize New Hampshires. They all died, nevertheless, and the royal chicken house was inherited by the local Department of Agriculture for use on special occasions.

Building D, refurbished when it came to FAO in the 1970s, is air conditioned throughout, to the comfort of Administration and the FAO/World Bank Cooperative Programme. In the older buildings, coolers have been set in the windows of offices designated for directors and above. Thus cosseted they are expected to observe *tenue de ville* and receive visitors with appropriate dignity. The rest of FAO can come to work in sports shirts. This has its logic – in contrast to the practice once observed at the ultra-British Galleface Hotel at Colombo. The dining room there had only fans, but guests were required to wear coats and ties; as a result, they sat red-faced and sweating throughout dinner. The bar, on the other hand, was air conditioned; yet there one could go in wearing an open-necked shirt.

A great bugbear for FAO staff, and a cause of a great deal of lost time for the organization, was the bank. Opening hours for banks in Rome fall entirely within FAO office hours. Continuing requirements to pay electricity, telephone, tax, and most other bills in cash and the unwisdom of holding large sums in one's house or on one's person mean frequent recourse to the bank. Opening a branch in the FAO buildings saved staff time going elsewhere; however, its procedures were still cumbersome and slow. The departure of a colleague from his office for an hour or so would be heralded by the comment 'I'll go and see how bad the queue is at the bank.' The great breakthrough was the institution of cash dispensers. While one still has to queue to deposit a cheque, cash can be obtained simply by pressing some buttons.

When I first joined FAO I was told that the door was locked at five minutes past nine in the morning. If one arrived later, one's name and reason for being late went into a book. This could go the director general. For a professional officer this would be undignified. So if it seemed likely that I would not be able to arrive before the doors closed, then I should phone in and ask for half a day's leave. This procedure was instituted by an American Director General to try to ensure that FAO staff got to work on time.

Locking the doors of a building to keep staff in has also been tried in some Mediterranean countries. Once I had to visit the Ministry of Agriculture in Athens. At 11am I found the doors locked – and three young women suckling babies on the steps. The regime then in force kept all staff inside for the established working hours unless they had official business elsewhere. Mothers could go out to feed babies brought to the door by maids or relatives, but must remain in sight.

DG Sen, with his Anglo-Indian Civil Service background, found such measures distasteful. The FAO buildings were opened to anyone not obviously a beggar. Ensuring that staff kept reasonable hours was left to their supervisors. Handbags and other items were occasionally reported stolen following this, but this could often be traced to a kleptomaniac colleague. The doors closed again with the onset of PLO assassinations and hijacking in the 1970s. People could go out freely; they then had to show a pass to come in again.

An initial and sometimes continuing disadvantage of Rome for many people is the Italian language. Those who are already familiar with some other Latin language are inclined to underrate this. It is fairly easy for them to reach a plateau of competence in Italian that will serve day to day needs. More difficult is the situation of those from Africa and Asia for whom the working knowledge of English, required of all FAO staff is already a second language. Help in negotiating the essentials – renting an apartment, buying a car – is available from FAO and its Staff Association. Usually there will be people from the same background already in Rome, at FAO, at the embassy, or with some other organization who will also help to ease cultural and linguistic difficulties and provide a sense of community.

Ladies around Rome with spare accommodation saw FAO staff as highly desirable potential tenants. Not only would they add to the tone of the establishment, there could also be additional benefits – access to small luxuries from the FAO commissary for example. They could, however, be disappointed. A marchesa boasting to her friends of a senior American FAO officer as her tenant received as her first gift from the commissary a packet of soap flakes. Another lady renting to a distinguished economist, later to be enobled in Britain, had to pay for roof repairs because of damage done by his family sunbathing there.

ROME'S IMPACT ON FAO

When FAO came to Rome there was immediate pressure on it to employ Italians. At one point DG Norris Dodd called in the appropriate Italian government official, told him that FAO could not cope with the flood of applications, and asked him to submit a list of ten individuals who were highly qualified and worthy of consideration for employment. The names on the list ranged from those of a distinguished engineer to a charwoman who worked for a member of parliament.

Almost any Italian contact, once it became clear that one was in FAO, began to assess the potential for influence. I suppose I had preference over others for my very special apartment because the owner, a marchese, thought I was a 'pezzo grosso' in FAO. She would expect to benefit through jobs for her 'nipotini' and her friends.

The numbers of Italians in FAO built up steadily, not so much by qualification, but by attrition. When FAO first came to Rome, not many Italians could meet its entry requirements. They lacked competence in English; they lacked the experience of tropical agricultural environments of the British, the Dutch, and the French. Italians who had lived and worked in Egypt had the best chances. They spoke English, French – often Arabic also. However, while foreigners often found first place among applications for FAO posts, they also tended to leave them. They moved away to gain new experience and promotion, or went back to their own country for family and career reasons. Rome was not their home, and it was unlikely to provide favourable careers for their children. The Italians, in contrast, saw no better place to go. Girls who came in as trainee secretaries acquired husbands, but stayed on working in FAO. They built up family ties, property holdings, and careers for their children. There was no advantage to be had in moving. So gradually, through seniority and familiarity with FAO procedures, the Italians moved up in FAO, especially in Administration. Often they complained that no Italian received a high post. In 1990 there was an Italian Assistant Director General; another based in Rome was the FAO Representative for Europe.

Too strong an Italian flavour can be bad for FAO. If they are predominantly in view – and much Italian is heard in the corridors – it risks losing credibility as an international organization. A very well-qualified American with project management experience in several countries, a European wife, and the speaker of several languages saw this as a reason for not wanting to work there. With so many Italians in the offices and corridors, he said, 'one cannot take FAO seriously'. IFAD has done better in this respect, keeping a more conspicuously international staff.

Together the increasing proportion of Italian staff and the Roman environment have had their impact on work performance. The morale of the organization underwent a change as Italian attitudes and habits became predominant. Staff members from other parts of the world accustomed to higher levels of effort often slow themselves down to remain popular with colleagues. In an organization as big as FAO with a headquarters staff of thousands it would be difficult, in any event, to keep them all productive. Supervisors with strong work ethic back-grounds would check that their staff arrived on time, ask where they had been if not available when needed, and complain if their output was low. Even they tended to relax with time, taking long lunches with visiting colleagues 'to maintain contact' and turning a blind eye to a vacant desk. Periodically there would come a general exhortation to staff to be at their desks, keep telephones covered, and work harder.

An earlier DG had tried removing seats from the bars where many people went for mid-morning and afternoon coffee, even closing them outside meal times. This led, however, to the perhaps more adverse impression on the visitor of a number of offices empty, then one full of people drinking tea. There would be an electric hot plate on the floor and white cups around with initials daubed on in red.

An established Italian attitude that crept into FAO was that toward the taking of sick leave. If one's contract of employment made provision for a specified number of days on full salary while off sick, then one should take them. Commonly heard, especially among people on routine jobs, was the statement 'I haven't taken all my sick leave yet.' A doctor who came from Sweden in the 1970s to head the FAO Medical Unit infuriated the staff union by issuing a report showing that FAO lost 20 per cent more working days through sick leave than the UN organizations in Geneva and that a substantial part of its staff were alcoholics. On the first point he was right. Doctors in Italy are known to be complaisant in issuing medical certificates. Extended cures at popular spas and mud bath treatments are still fashionable in Italy. While the FAO medical insurance did not cover the residential charges

112

involved, Italian doctors were generally prepared to issue certificates covering the time off work. Less job security in the UN at Geneva was also a factor in the difference between the two organizations. The Swedish doctor could not hope to abolish these practices overnight. He hoped, however, to reduce their incidence by arranging for follow-up visits to staff who were frequently on sick leave.

Reuter's news item on this report also noted that 30 per cent more staff in FAO incurred psychiatric problems than in other UN organizations. These are not the kind of grounds for sick leave that the typical Italian would present to his or her doctor, however complaisant. It is an easy guess that the bulk of the staff affected were foreign women. Commonly attributed as the cause was the unsatisfactory nature of their relationships with men. Did the Roman environment stir up a special desire to belong that could rarely be achieved, or was it just that more of them than in the other organizations were away from their home country?

On the issue of alcoholism, FAO contentedly maintained that the Swedish doctor's own background led him astray. As his own assistant put it, 'By his standards the whole population of France and Italy were confirmed alcoholics.' In these countries aperitifs and wine are part of a meal; taken with food, alcohol has much less impact. This doctor did not last very long in FAO.

The Italian tradition of giving a present to the boss or to persons of influence also came into FAO. Some general service Italians handed over a part of their salary every month to the person who got them their job or kept them in it. This died away as people became more secure. An English secretary recalls, however, going into a supervisor's office with a thick, typed report bound with string. Only half looking up he said, 'Oh is that a present for my son's birthday? Put it over there with the others.'

While located in Italy, FAO did not presume that its host country needed active professional assistance. It was a volunteer team of FAO staff that undertook a relief mission to Capranica, a mountain village near Rome that had been cut off by heavy snow. The local priest had called for help on a short wave radio. There were people without food, he said, and a man in urgent need of an operation. We trod a way through deep snow carrying food on our shoulders. On arrival we found the village square full of men standing around doing nothing. 'Why did they not clear the road?' someone asked. 'Who would pay us?' was the answer.

FAO'S IMPACT ON ROME

While the Italian influence on FAO has been steadily pervasive, the reverse impact can only be called modest. FAO arrived as a source of jobs and material goods when Italy was still poor. Then came its economic miracle. Expanding new industry, new commercial enterprise, the dominance of the fashion market – these counted for much more than FAO. From pleading for 'recommendations' for their daughters for a post in FAO, Romans shifted to informing one that a better-paid post had been obtained for them in commercial business.

Even so, FAO has provided direct employment for well over a thousand Italians and indirect employment for thousands more. FAO international staff have rented and bought houses, spent money in hotels and restaurants, and employed maids when they could get them. FAO conferences and meetings have brought in annually thousands more visitors on expense accounts and representation allowances. Swissair is said to have maintained its extensive Third World network largely on the United Nations business with Geneva. FAO business must also have contributed substantially to the income of Al Italia, which generally offers the most convenient flights in and out of Rome.

Individually, however, FAO people are not big spenders. Indeed, it was said that they bargained over prices more strenuously than the locals. They have been told that they would be asked prices higher than normal and should seek discounts on everything. Lists of shops and restaurants were circulated that would give special prices to FAO staff. They had access to the catalogues of Sears Roeback and Quelle, and used these sources for goods that were priced high in Rome. There was also the FAO commissary.

For many FAO people the salary they received from the organization was much higher than that on which they had lived before, and they tended to maintain their previous tastes and economical habits. There was never a senior officer who attracted attention by a lavish mode of living, or the possession of a luxury home with a retinue of servants. The bulk of them were saving for when the job finished, for when they went home. In FAO, snobbery could be inverse. Some of the highest placed officers competed verbally as to which had come from the humblest beginning. The Americans boasted of poor farms in the south. In contrast, many Italian women possess not one fur coat, but several. They dripped 18 carat gold. Middle-class Italian families had several houses; an apartment in town, one by the sea, another in the mountains. In restaurants they ordered lavish meals for the 'bella figura'. Behind all this, however, according to the statistics, the Italians as a whole were also great savers.

Surprising, since FAO was a world institution conspicuously situated in the middle of Rome, has been its continuing low profile with the Italian media. Two weeks after prime minister Andreotti had opened the 1989 FAO Conference there was a quiz programme on Italian television. A group of children were asked to name an organization beginning with the letter F concerned with food and agriculture. None could do it. The presence of a senior member of the Italian government at the FAO governing conference brought notice in the press and on television, but not the conference's business. The institution of World Food Day on 16 October annually was a good move in this respect. It provided a recurring event to which some information could be tied.

Rumours that FAO might be leaving Rome caught the press, of course, and very occasionally its issues with the staff union. The substance of its work did not. Only when the leader of the Italian Radical Party, Marco Pannella, fasted to secure a major Italian programme of assistance to the developing countries did this subject receive major attention. Italy then became one of the leading aid donors of the 1980s.

Some people said FAO's poor publicity was a communication failure on the part of the Information Division. It could, however, place FAO news items and recorded talks with Britain's BBC, which won an FAO prize for its co-operation. Though there were good Italians on its staff, FAO Information never made much impact in Italy. In 1972 Cervi of FAO was elected President of the International Union of Agricultural Journalists. The retiring president, F.P. Couvreur, Chief Editor of the *Journal de la France Agricole*, said Cervi's election was not only an act of appreciation of his personal services to journalism, but also of the value of the work of FAO's Information Division. A sardonic comment would be that from experience close at hand, the average Italian was sceptical as to the likelihood that a public body would do anything very effective. In 1980 an Italian was appointed Assistant Director General in FAO with responsibility *inter alia* for its Information Service. Was he expected to strengthen FAO's public image in Italy, or was it part of a settlement for that country's vote at the last election of the director general?

FAO people have made individual contributions to the small print columns in the Rome newspapers. But this also has been modest. Divorces are hardly noted, nor errant daughters; the suicides are generally masked.

There are Italians also who see FAO's technical potential. One day F.P. Pansiot, an FAO horticulturist, received a message that the director of the Rome opera wanted to meet him. Pansiot was quite excited: he liked good music and wondered what aspect his visitor would wish to discuss. It turned out that the director was growing mushrooms in the vaults of the Terme de Caracalla where operas are staged in summer. He wanted advice on the most suitable kind of manure.

An Italian who appreciated FAO in depth was Giordano Dell'Amore. He began with the Cassa di Risparmio della Provincia Lombarda as an usher and ended up as President. It had grown rich in the richest part of Italy, financing the city authority of Milan until its taxes came in. Operating under a law carried over from the Austrians it had few taxes to pay itself and no shareholders seeking dividends. Its 'surpluses' could be used for good causes – helping the Milan hospital, the Boccone University, making up the deficit at the Scala opera house. Having done this for a number of years, Dell'Amore thought he could look further afield. He approached FAO for collaboration in a study to determine the main needs for assistance to Africa in credit for agriculture – the basis of most African economies. He provided the finance, but accepted FAO's management, seeking only that some of his staff learned by participation. Out of this study came two leads that he backed with substantial resources. One was the scope for promoting banks that would attract rural savings and lend them back to agriculture. The prevailing commercial banks and the post office savings systems in Africa all tended to channel rural money to the capital city. In several countries – Ghana and Somalia, for example – rural saving and lending banks were established with his support. The second lead was the need for intensive training of staff for rural banking. Dell'Amore set up 'Finafrica' on the outskirts of Milan as a continuing training institution, and provided scholarships for Africans to study and gain experience there.

Dell'Amore backed the World Credit Conference held at FAO in 1976 with hospitality and participation. FAO was duly grateful for his efforts – not so his country. Along with other bankers, he was prosecuted by a politically oriented magistrate. He was arrested at dawn and dragged off to jail. The charge was dismissed, but Dell'Amore was well into his seventies and died shortly afterwards.

There was also a marchese who wanted to give to FAO, for use as a training centre, a near-deserted village in Tuscany. There were abandoned houses that could be used for accommodation, a hall suitable for group lectures and meetings, and farm land around that

116

could be used for practical instruction. FAO sent a delegation to assess its potential. Friendly words were exchanged at a dinner by the sea, but the response had to be negative. For people from the developing countries the village afforded no advantages over training in their own environment and would be much more expensive.

In Latin American circles FAO representatives have often been teased about the bull. Its owner complained that it kept eating, but did not perform. A neighbouring farmer commented it must be an FAO bull; it was only there in an advisory capacity.

Part 2

POVERTY ALLEVIATION: FAO'S PERFORMANCE IN THE AREA OF MARKETING

8

MARKETING AND THE RURAL POOR

The demarcation point for poverty in the World Bank's Development Report for 1990 was an annual income of $370. The number of people below this was 1.1 billion. One-third of these were in India. By this definition, half of the people in Africa south of the Sahara were below the poverty line, as were most of those in north-east Brazil and the Latin American mountain zones with high Indian populations.

THE DIMENSIONS OF POVERTY

Defining poverty in money terms begs many questions: the value to be set on a house and subsistence production; the cost of the essentials that must be purchased in the country concerned; the exchange rate applicable, and so on. For Professor Vyas of India it is a food intake averaging not more than 2,100 calories per day, with associated low access to health services, housing, and water (IFAD 1990). In the Mexican states of Oaxaca and Chiapas, calorie consumption averaged 1,500 and 1,600 per day, respectively.

There are many places where average income is far below the $370 level. In 1990 one-third of the population of Ethiopia were at $120 or below. In Peru the poorest 50 per cent of the population were also below this level. As a reference point, $1.30 to $1.50 per day was the wage paid to workers in commercial agriculture in India, Kenya, and similar countries with an ample labour supply. They would also possess a house and some land for subsistence cropping.

Poverty need not equate with unhappiness. The bulk of the people of Malawi see their low level of living as normal. Tourist and other displays of greater wealth are not common there. Poverty for Malawians shows up in low infant survival rates and family malnutrition in areas where farms are small and crop yields low.

121

The World Bank's Development Report 1990 used 1985 figures. Since then the situation has deteriorated because of:

- population increase – more people need food and employment;
- economic recession with lower prices for exports, reduced capital inflows, high indebtedness;
- structural adjustment programmes resulting in reduced government expenditures and higher prices for food that have impinged negatively on those who were not producers;
- successive years of drought, and continuing civil war in a number of countries.

In south-east Asia most national economies continued to grow throughout the 1980s. There, the poor were groups that had been left behind because resources and opportunities were limited in the areas where they lived. In much of Latin America, during the same period overall income fell back sharply. In Africa, south of the Sahara, incomes fell, as did food output per person. Populations, however, continued to grow – in some countries at an alarming pace. Growth rates of 4 per cent per year prevailed in North African countries, making them increasingly dependent on food imports. Kenya, also with a 4 per cent annual increase in population, will have increasing difficulty in providing lasting food security for its people.

Muslim fundamentalist pressures in North Africa and the fragility of some other African governments blocked effective promotion of birth control where it was needed. In China and India there have been very determined efforts to limit population growth; in recent years these have been relaxed.

Historically, the way out for the poor in a disaster-prone area, or one over-populated in relation to its resources, has been to emigrate. Many workers on cocoa farms in Ghana and the Ivory Coast came from the Sahel. People from the Sahel also went to Sudan when irrigated cotton-growing in the Gezirch set a new demand for labour. Such movements continue. It has been said that if research on crops for dry land agriculture does not make a breakthrough soon, most people will have left it.

There are some Third World countries with small populations and substantial income from petroleum resources. Here governments have been able to ease the life of the poor, providing services and cash allowances. Other countries are aware of a considerable disparity of individual incomes, but see the tax base as too small and the poverty sector too large for significant income transfers to be practicable. Even so, they could attempt to extend basic services – primary education and low cost health support, for example. Some of the services they do provide, such as the distribution of low cost food grains, are poorly targeted. They go to government personnel and the urban middle-class, neglecting the rural areas where 80 per cent of the poor live. That more is not done for the rural poor, including a shift in the relative burden of taxes from agriculture to urban activities, has to be attributed to political preference for the status quo.

'Food for work' programmes can be organized in rural areas to relieve situations of extreme stress and to engage labour seasonally without employment in the building of public infrastructures. Lasting solutions to rural poverty, however, require that the people be enabled to earn more, or support themselves better, on a self-sustaining basis. This was the conclusion of the 1979 FAO World Conference on Agrarian Reform and Rural Development which required the FAO to report on follow-up action (FAO 1988). In general terms, it recommended:

1 Improved access to land, and irrigation to make it more productive.
2 Better access to inputs, markets, and related services.
3 Provision of advice and training on improved agricultural production methods.
4 Development of non-farm rural activities.

The Conference also stressed people's participation and the role of women.

The recommendations constitute a very broad agenda. They correspond to the responsibilities of several major divisions in FAO. To deal with them as a whole risks being administratively formal or professionally superficial. Here we shall concentrate on the second heading – access to commercial inputs, particularly fertilizers, and to markets. This is the area with which the author has been directly concerned. Off-farm activities to provide additional employment will also be treated in parallel. Provision of fertilizers and other farm inputs, the transport and processing of farm output, and handicrafts using local materials are likely to provide the more immediate work opportunities in poor rural areas.

THE ROLE OF MARKETING

Marketing serves equally the poor, the intermediately well-off, and the already rich. Even the subsistence smallholder must sell some produce if he is to have the cash to pay for inputs and services that will raise his level of living.

In accordance with current academic practice, marketing is defined as the business activities associated with the flow of goods and services from production to consumption. The marketing of agricultural products begins on the farm, with the planning of production to meet specific demands and market prospects. It is completed with the sale of the fresh or processed product to consumers, or to manufacturers in the case of raw materials for industry. Agricultural marketing also includes the supply, to farmers, of fertilizers and other inputs for production. Marketing has a parallel role in the development of commercial fishing.

The tasks and responsibilities of marketing may be summarized as follows:

1   Finding a buyer and transferring ownership.
2   Assembling, transporting, and storing.
3   Sorting, packing, and processing.
4   Providing the finance for marketing and risk-taking.
5   Assorting and presenting to consumers.

If marketing is to fulfil its role of stimulating and extending development, specific enterprises must be responsible for finding foreign or domestic buyers for the various types and qualities of produce. They must be able to arrange assembly from farms; packing and presentation in appropriate containers; sorting according to buyers' requirements; transport to buyers' depots or markets which they attend; storage to extend the availability of seasonal commodities; and processing to extend the time and range of sales outlets. The enterprises must provide the necessary investment capital for fixed facilities, and the working capital to carry purchases from farmers until resale proceeds are received. Implicitly, these enterprises must possess the financial resources, the qualified managerial, sales, and technical personnel, together with the initiative and willingness to accept business risks, which are necessary in order to perform these tasks efficiently. In export marketing, or in substitution for imports on domestic markets, they must be able to match the competence of rival enterprises in other countries.

Marketing enables a person with some land to move from semi-subsistence to growing produce regularly for sale. Correspondingly, it allows an increasing proportion of a country's population to live in cities and buy their food nearby. Marketing also provides an incentive to farmers to grow produce for export. This increases farmer's income so that farmers form a growing market for domestic industry as well as earning foreign exchange to pay for essential imports.

An efficient marketing sector does not merely link sellers and buyers and react to the current situation of supply and demand. It also has a dynamic role to play in stimulating output and consumption, the essentials of economic development. On the one hand, it creates and activates new demands by improving and transforming farm products and by seeking and stimulating new customers and new needs; on the other hand, it guides farmers toward new production opportunities and encourages innovation and improvement in response to demand and prices. Its dynamic functions are thus of primary importance in promoting economic activity and creating employment. For this reason an efficient marketing sector has been described as the most important multiplier of economic development (Drucker 1958).

Some illustrations of the benefits to semi-subsistence farmers, fishermen, and landless workers that accrue from effective marketing initiatives follow.

Tea processing plants run by the Kenya Tea Development Authority have provided a profitable outlet for 138,000 small growers. These smallholders obtained cash incomes well above the average for their area. In a similar manner, 28,000 growers and associated workers for the Mumias Sugar Company in Keyna earned improved incomes. For a number of crops, smallholder production is advantageous because the incentive for care in harvesting is direct. This is the case with tea in Kenya and marigolds in Ecuador, also tobacco in various countries. Smallholder production of sugar is now preferred because of the risk that organized labour on large plantations would disrupt harvesting at the time of optimum yield.

People with no land of their own have been brought into profitable poultry raising systems. The Charoen Pokphan enterprise in Thailand saw a profitable market in Japan for boned-out chicken. It furnished young chicks together with the feed, drugs, and veterinarian services required to raise them. Those who had no chicken houses could build them with bank loans which it negotiated. Large numbers of women were employed in dressing and packing plants.

Thousands of cattle owners have benefited from small private and larger cooperative milk marketing developments in India, and from livestock marketing initiatives that have expanded outlets for traditional suppliers in Botswana, Chad, and Swaziland. Continuing access to the favourable market provided by the Botswana Meat Commission has stimulated a livestock offtake much higher than on similar range-grazing conditions in other parts of Africa. It has been of broad and immediate benefit to the low income livestock-dependent rural population.

Up to 1960, shrimps caught by Indian fishermen were discarded or spread around coconut trees as fertilizer. In 1980, exported frozen, they earned $45 million from US markets alone.

Employment generation is a core factor in alleviating poverty. In addition to stimulating production that gives employment, agricultural and fish marketing enterprises also employ large numbers of people directly themselves. The Asian Productivity Organization estimated the annual rate of growth of employment in agricultural and marine products processing 1970–75 in developing countries at 7.9 per cent. This is much higher than the rate of population increase for the period (about 2.8 per cent). Employment in processing often includes paid work for women; they make up 60 per cent of the employees in tobacco marketing in India. Large numbers of women are also put to work in the picking and marketing of fruit and vegetables needing careful handling to obtain the best price.

MARKETING ENTERPRISES

The enterprises that constitute the driving force in marketing may be independent individuals and partnerships, joint stock companies based locally or in another country, co-operatives and marketing boards, corporations or authorities set up by government.

The poorer producers tend to incur higher transaction costs than larger units because the quantities of inputs they need, and of output they sell, are smaller. They are often less well informed and have less bargaining power. Important for them are the consideration and assistance they receive from a marketing enterprise. Will they be treated fairly at the local buying stage? Will they be advised what to grow and when, and on how to raise its market value? Will they be helped with credit, and with access to the inputs they need? FAO compared the various forms of marketing enterprises in these terms (see Table 8.1). It summarizes the services they are likely to offer the smallholder, their limitations, and the kind of government support needed to raise their effectiveness.

Private marketing enterprises

These have shown themselves able to start up and go a long way with very little capital. They are great builders of capital assets. Their operators tend to be economical, even parsimonious, in their personal expenditure, very careful in their business outlays, stringent in their requirements of performance from paid staff. They are able to operate at very low cost. Only those staff who make a positive contribution to the enterprise are employed. Full use is made of family labour. Outlays on equipment and other capital expenditure are kept to the minimum.

The structure of private marketing enterprises has the important advantage of continuity without drain on public resources. Many private marketing enterprises help regular farmer clients with practical advice on market requirements and input use, and are the most easily accessible source of credit. The private trader assumes risks over resale prices and loan repayment against which he tries to protect himself through a target margin. To the farmer this often seems high. His protection against excessive margins is the ability to go to another trader. So there must be competition, and the main role of government in this regard is to maintain competition by establishing markets where many traders will be present, prices can be compared, and information provided. Governments will also be called upon to moderate extreme price fluctuations where practicable.

Because of flexibility in decision making, private marketing enterprises have shown themselves especially well suited to handle perishable produce, livestock, meat, and eggs, and to see new opportunities for marketing initiatives. Because of their local knowledge and use of family labour they can combine economically the purchase of farm output, sale of farm inputs and of consumer goods on the small scale, adapted to specific rural areas.

Transnational marketing and processing

Transnational marketing enterprises have helped the small-scale farmer substantially, enabling him to participate in intensive production/ marketing contract systems. He is assured a market outlet, and he receives a full set of services on credit. The quality of the extension assistance provided far exceeds that normally available because it is tailored to the needs of the market outlet served and the processes used. It is likely to be based on specific research and be backed up by the direct provision of seeds, pesticides, and fertilizer on credit, combined

Table 8.1 Relative advantages for small-scale farmers of alternative marketing structures

Marketing structures	Sales position of small-scale farmers vis-à-vis larger-scale farmers	Sales position of small-scale farmers vis-à-vis buyer	Technical assistance	Seed/planning materials provision	Fertilizer supply	Other credit	Government support required
Independent private marketing enterprises	Bargaining weight can be used against small farmer	Advantage of access to alternative outlets	Advice based on local experience	May supply on credit	May supply on credit	Consumption credit often available in addition	Provision of market infrastructure and information services; maintenance of competition, some price stabilization
Transnational marketing and processing	Equitable prices if it can make a contract	Dependent but can secure if can meet product quality requirements	Can be direct and intensive	Direct supply on credit	Direct supply on credit	No	Should negotiate participation for small-scale farmers and prices

Co-operatives	Equal if operates successfully	Favourable provided co-operative operates efficiently	Collaborate with government service	May arrange a supply	Direct supply on credit	May supply where supported by co-operative bank	Continuing financial support, supervision and protection generally needed
Marketing boards/state trading agencies	Equal price if can reach official buying station	Depends on access to buying station; may be subject to illicit charges	Usually left to government service	Assistance rare	Assistance rare	No	Insistence on measures to help small-scale farmers at rural buying points
Development companies/authorities (parastatals)	Equal prices	Protected provided can meet product quality requirements	Can be direct and intensive	Direct supply on credit	Direct supply on credit	No	Major financial input or privileges usually required

with day to day advice on how, and when, to carry out production operations and the harvesting and handling of the product.

Typically, this kind of marketing service has been initiated for export crops by companies like British American Tobacco Company for tobacco, and the cotton companies in francophone Africa. It has been extended to sugar, tomatoes, and other crops for domestic markets. Even so, only a small proportion of farmers can expect to obtain contracts with such enterprises.

The disadvantage is the potential for captive status. With a farming pattern adapted to the buyer's requirements the grower loses bargaining power over prices and quality determination. In Taiwan, Thailand, and elsewhere mediation by a government department has been needed for an equitable agreement.

Co-operatives

Co-operatives, or farmers' associations, can enable small-scale farmers to economize on transport to distant outlets and increase their bargaining power in negotiating terms of sale. Their success depends greatly on the managerial capacity that can be generated locally and on the cohesion of the group in face of opportunities for individuals to benefit by going outside it.

In principle, co-operatives constitute a very favourable instrument for improving small-scale farmers' bargaining power on the market and channelling new inputs and technologies toward him. In practice, co-operatives in developing countries have done best in assembling not very perishable crops such as coffee and cotton. In India and Kenya they have managed the regular routine of the dairy. In most countries they have needed special protection and substantial government financial and administrative assistance. Conditions favouring co-operative or other group marketing arrangements are:

- specialized producing areas distant from their major markets;
- concentration upon and homogeneity of farm production for market;
- groups of farmers dependent on one or a few crops for their total income;
- availability of local leadership and management;
- an educated membership;
- members with strong kinship or religious ties.

Intrinsic handicaps of the formal co-operative for very small farmers are:

lack of capital, if each member is to contribute an equal share; difficulty in finding competent, motivated managers; and politicization. With their bureaucratic procedures, most co-operatives have difficulty in competing with private traders in prices and services to small-scale farmers.

Development authorities

In an effort to ensure that small-scale farmers, in particular, received the services they needed, many governments embarked on integrated development programmes during the 1970s. With aid agency support they set up development authorities that would provide market, input supply, credit, and technical assistance services to the farmers in a designated area. Often they were land settlement or reform projects. Sale of farmers' produce through the authority facilitated recovery of land payments and credit. The interests of the small farmers were cared for under such an authority, if it was managed well. The intention was that models tried out on a single area basis should be replicated over a country as a whole. This did not go very far because of the recurrent costs of such bodies in the face of declining government revenues during the 1980s.

Marketing boards/state trading agencies

Through the 1960s, marketing boards assigned an export monopoly over certain products by government action demonstrated ability to carry out major assembly and sales operations. They standardized quality and packaging and achieved benefits of scale in transport, processing, and sales. Farmers could count on receiving a pre-announced minimum price if they delivered their produce to the board's buying station or agent. Usually this price was cushioned against sharp fluctuations on international markets by a reserve fund. The defect of this system was that the prices paid to producers tended to decline as a proportion of the export price in real terms. Exchange rates were maintained at levels adverse to exports; the export board was a convenient mechanism for tax collection; more was paid into reserve funds than was paid out to growers. In Ghana, for example, this led to massive smuggling of cocoa into neighbouring countries and to declining production.

Relieved of these adverse features, a parastatal can still be helpful to small producers. As an alternative to highly fragmented export marketing channels, the parastatal Interbras exploited the advantages of scale in Brazil without needing a monopoly. Lobster fishermen with only one or two boats sold to Interbras directly. It sent the lobsters frozen to the USA. Controlling the bulk of the supply from Brazil, it used cold storage to manage the market there and obtain better prices.

Parastatals to stabilize supplies and prices of food grains on domestic markets followed the export model. They bought grains and other products at pre-announced prices when offered at their buying depots or to their agents. Maintaining a minimum price in an otherwise free market, they have been a valuable protection to farmers. However, where operated as a monopoly to enable them to cover the costs of pan-territorial pricing and of holding reserve stocks, their prices have often been a disincentive to production.

The question that must be raised is whether a marketing board set up primarily to concentrate export bargaining power or to stabilize prices will concern itself very much with services to small-scale farmers. Their buying unit staff frequently require tips before they will give farmers attention. Several such boards have declined firmly to enter into the provision of fertilizers or the handling of credit repayments. Many have little concern for what happens before produce arrives at the buying station, leaving the small producer to pay for transport or depend on an intermediary.

Parastatal organizations designed specifically to serve smallholders and the landless have been tried in various countries. They have the best chance of viability where: (a) the small farmers are concentrated in a particular area; and (b) the farmers offer produce in which they have a cost advantage.

This is the logic behind the establishment of a 'Co-operative Corporation' in the Andhra Pradesh tribal area of India. It bought other produce and sold inputs and consumer goods, but the handling of forest products brought in by the tribals was its lifeline. While requiring government capitalization and subsidies, it was near to breaking even on current operations in 1989. Its prices to tribal suppliers were claimed to be 30 per cent higher than those offered by private traders in areas where it had no buying point. Where smallholders live amongst and grow the same crops as larger farmers, assistance through a special enterprise is more difficult. The Ghana Food Distribution Company is supposed to buy directly from small farmers, so it buys only small-sized lots. In practice, it has been buying a small part of various farmers' marketable surplus.

Competition between private marketing enterprises and government-supported parastatals operating alongside has brought the average marketing margin on food grains in some South-East Asian countries down to 20 per cent of the consumer price (see Table 8.2). The much higher margins in the African countries reflect higher transport costs. That they were higher in Kenya, Malawi, and Tanzania than in Nigeria and Sudan can only be explained by parastatal monopoly pricing and the extra cost to private enterprise of evading it.

Table 8.2 Average marketing margins as percentages of prices paid by consumers: selected African and Asian countries 1975–80

Country	Product	Producer as % of consumer price	Marketing margin as % of consumer price
Tanzania	maize	38.2	61.8
	rice	56.6	43.4
	sorghum	48.1	51.9
Kenya	maize	42.0	58.0
Malawi	maize	48.2	51.8
	rice	55.1	44.9
Nigeria	maize	54.5	45.5
	rice	57.0	43.0
	sorghum	59.8	40.2
Sudan	sorghum	61.2	38.8
Bangladesh	rice	79.0	21.0
India	rice	82.0	18.0
	wheat	79.5	20.5
	sorghum	80.0	20.0
Philippines	rice	87.0	13.0
	maize	71.5	28.5
Indonesia	rice	84.0	16.0

Source: R. Ahmed and N. Rustagi (1985) 'Agricultural marketing and price incentives. A comparative study of African and Asian countries', Washington: IFPRI.

PHYSICAL INFRASTRUCTURE FOR MARKETING

Essential if marketing initiative is to have an impact on a poor, under-developed area are roads, transport vehicles, storage and distribution facilities, processing technology and materials, and means of communication and access to information.

Many places are cut off from sources of fertilizer and markets for their produce during certain seasons; there are many more still where the cost of transport over bad roads, and the time it takes, inhibit marketing of the more valuable perishable products. A study in Ethiopia showed truck charges over earth roads with deep gullies and wet spots were five times more than those over tarmac. In Zimbabwe, with a relatively good road system, the limit to lower-cost group purchasing of fertilizers is set by accessibility with a loaded vehicle. Tarring a main road going into a project area and opening up, or rehabilitating, feeder roads has led to reductions in marketing costs of up to 60 per cent and the use of fertilizer on holdings that were previously inaccessible.

Storage

Storing grain in their own home, or in a special structure nearby, is most convenient for smallholders, as it can be supervised directly. Participation in group storage arrangements involves transport to and from the store, and the contentious issue of assuring to each participant that the grain he takes out is equivalent in quality to what he put in. Use of insect-resistant varieties of maize, rice, or cowpeas reduces storage losses; family and other consumers in the production area may also prefer them. Sachets of insecticide powder convenient for smallholders are available, but involve a cash outlay that may not always seem justified.

The main purpose of storage, in addition to conserving supplies for family use later in the year, is to avoid having to sell at the low prices usually prevailing just after harvest. Seasonal price movements for maize in Ghana are shown in Figure 3. Wholesale prices of paddy in the Madras area of India rise seasonally by about 30 per cent. A post-harvest price increase of 5 to 17 per cent is needed to cover the cost of grain storage, depending on its duration and the interest rate on finance.

A first recommendation on a proposal to invest in expensive refrigerated storage for perishables would be to look for changes in production and marketing that would reduce the need for it, i.e. extend the harvesting season by planting early and late varieties, or the use of plastic to promote early growth or late maturity. The figures below, derived from FAO experience in the Near East, show that storage costs can absorb much of a wide seasonal range in price. These figures refer to a price difference of $60 per ton between the time of picking apples and

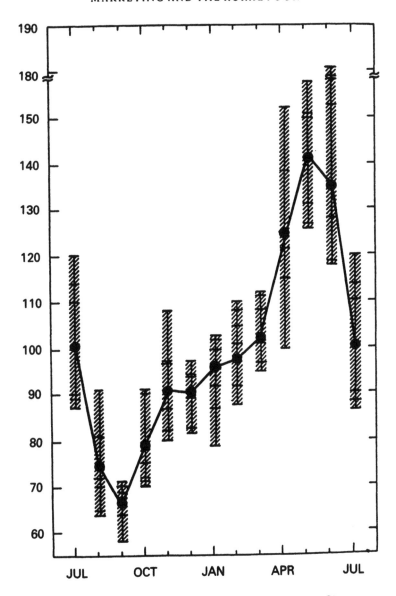

Figure 3 Seasonal price movement for maize, Ateburu, Ghana
1965–74

Source: Ghana, Ministry of Agriculture, *Monthly Food Situation Report*, various issues.
Horizontal bars show prices in individual years, large dots show monthly averages for the
period.

135

six months later. The profit margin in storing until then was only $4.70 per ton.

	$ per ton
Estimated increase in price over 6 months	60.00
Storage costs:	
rent (or overhead cost if self-owned)	21.00
storage losses (weight and quality)	23.80
rent of crates	4.30
interest (10%) on initial value of stocks	6.20
Total costs	55.30
Net margin	4.70

Processing

The use of suitable processing technologies can greatly extend the market for farm and fishery products. It can reduce waste, improve palatability, prevent spoilage and simplify handling and transport. It can adapt products to the needs and preferences of consumers in distant places.

Well suited to small-scale operations are the husking of paddy and grinding of wheat and maize for local consumption, the squeezing of oil from palm fruit, the processing of cassava to eliminate the bitter element and offering it for sale ready to eat as gari, the drying of tomatoes, drying and grinding of pepper, the drying and smoking of fish, the processing of livestock into various forms of meat, hides and skins and of milk into a range of dairy products. Many of these have special appeal that goes beyond purely local customers. Hygienically presented and packaged they can be marketed successfully to higher income consumers. Use of an attractive traditional package will add to the appeal. The preparation and marketing of appealing processed products may be the fastest way of increasing smallholder incomes.

Caution is urged in proposing the adoption of equipment developed for different conditions. It was recognized early in a Nigerian fisheries project that the standard smoking kilns available might not be suitable for the people on the project. The first step would be to test them out with small groups of women, and modify them according to the comments received.

Distribution facilities

Small-scale assembly, wholesaling, and retailing of farm and fishery produce can begin from the operator's house. It involves no additional cost, minimizes time spent going to and from work, and facilitates the part-time use of family labour.

Sale at an organized local market has the advantage of access to alternative buyers and to information on current prices through enquiry at a number of sources. Table 8.3 presents information obtained by an FAO-assisted survey of the smallest farmers in Kenya. It shows that a large proportion of them sold their produce at rural markets – though for many products guaranteed prices could be obtained from a marketing board. However, less transport was involved in a trip to the market, and for some varieties, at certain seasons, and for produce sold retail, the price at the market would also be higher.

Such markets should be within reach by foot or bullock cart, i.e. within 12 km distance. The outlay on fixed facilities can be low, using local materials. Important is ease of access under all weather conditions,

Table 8.3 Smallholder sales by place: Kenya

	On holding	Rural market	Roadside	Processing plant, milk collection centre, etc.
	Percentage of total			
Maize	26	45	8	21
Beans	22	57	11	10
Finger millet and sorghum	6	90	4	—
Sugar cane	39	55	3	3
Potatoes	47	51	2	—
Cabbages	44	48	7	1
Tomatoes	5	86	8	1
Bananas	32	64	4	—
Chickens	37	48	5	10
Eggs	14	71	10	5
Milk	21	7	28	44
Sheep	49	50	—	1
Goats	39	61	—	—
Cattle	46	43	3	8

Source: Integrated Rural Survey, 1975

protection of market users against theft and harassment, and the provision of relevant information on prices. The extension of rural market networks is specifically helpful to the smaller growers. The larger farmer can more easily justify transport to an outlet that is further away.

Assembly markets attended by wholesalers with access to transport expedite sales to more distant outlets. Location on a trunk road and good communications, preferably by telephone, are the essentials. At Asseswa in Ghana, large quantities of maize are assembled on the ground. Two market centres serve the mountain producers of European-type vegetables in Java. The only physical facility is a blackboard showing prices reported by telephone from Djakarta and other cities. Growers meet wholesalers there in the early morning. When a sale is agreed they go together with a truck to a place where the produce is stacked ready by the roadside.

Wholesale markets for fishermen are located at convenient landing points with a good road to consuming centres. Paved areas where fish can be displayed, shaded against the sun, with running water for washing them down, are the basic facilities. Sales proceed by negotiation, or by auction where the quantities offered and the number of buyers justify it.

Central wholesale markets develop primarily to serve urban retailers of perishable produce needing fresh supplies regularly, and should be located within easy reach with the transport vehicles they use. Ease of access from the main producing areas is another important consideration. Small growers at a distance usually send produce individually or in groups to commission agents. Where space is available they may also be allotted areas from which to sell directly.

The environs of such markets, and of rail and bus stations, are also favoured by migrants from rural areas selling produce retail. As mobile pedlars evading market charges, they hope to make a small margin on their capital outlay or credit allocation. Local government and market administrations often have a negative attitude toward informal traders. They are denied sales opportunities, and moved on by the police. However, beginning as petty retailers they earn some income when otherwise they would be unemployed. Some of them may build up the capital and 'know-how' to enable them to undertake extended operations.

Essential for effective marketing, also at the smallholder level, are convenient means of communication and access to reliable information. The substantial differences in prices for similar produce within countries like Nigeria reflect delays in the arrival of reliable information that could be the basis of arbitrage transaction. Use of standard weights, measures, and quality specifications that are generally understood helps to make sales by telephone or letter possible, without the need for a personal visit to examine each lot. Uncertainty over the amount and quality of produce offered impedes sales over long distances and leads buyers to discount their offers.

That regular traders obtain information by word of mouth or over the telephone, while irregular market participants – especially the smaller farmers and consumers – have no such channel, seems to load price bargaining against irregular traders. Insufficient allowance is sometimes made for the effect of competition in generalizing the benefits from information received by private communication. Even so, market transparency can be improved greatly with the issuance of crop forecast data, setting up blackboards in assembly markets showing the prices of the previous day and those at other relevant markets, and regular reporting on the radio.

It is said of various countries that during the 1960s and 1970s the infrastructure for marketing was good, but government policies were adverse. In the 1980s policies improved, but achieved only modest results because the infrastructure had become defective. In much of Africa, port handling was delayed, electric light and power uncertain, and spare parts and fuel for transport and processing were lacking. Not only do such uncertainties and difficulties add to the costs of marketing produce and obtaining supplies, they can also reach the point of discouraging production. The small farmer who sees his produce deteriorating for lack of transport to a market does not easily lay out funds for production the following year. Those growing perishables at a distance from their market are especially vulnerable. The Kigezi Vegetable Growers' Co-operative Federation, with over 4,000 small farmer members, collapsed when regular transport to Kampala could no longer be assured.

Because of transport and access uncertainties, land-locked countries like Malawi incur double, and even triple, the normal costs on imported fertilizers. In addition there are the costs of holding large stocks to be sure that supplies are available when needed. Some of these extra costs are beyond the responsibility of national governments. Unfortunately, the impact on farmers and fishermen of these delays and difficulties does not always receive the recognition in central government circles that it merits. Priority in the allocation of the foreign exchange that is available does not go to the provision of fuel, of transport vehicles and of spare parts to keep them running. Nor, in many countries, is enough attention given to equipping local repair services with the resources and materials needed for adequate maintenance.

INSTITUTIONAL AND POLICY CONSTRAINTS

Many of the most conspicuous constraints on marketing by small farmers and fishermen flow from the infrastructural inadequacies set out above. The timely movement of crops or arrival of inputs can be blocked by roads that become impassable or by a truck that cannot be moved. Produce cannot be protected against humidity, materials needed for processing are not available, information on imports or other events that will affect prices on smallholders' markets is not passed on.

Marketing initiatives can also be inhibited by specific government directives. For two decades, many Third World countries were led by men who scorned private enterprise and tried to force development into a socialist mould. State marketing boards and official co-operatives had full support; other marketing was restricted, impeded, and starved of resources.

The elements of an economic environment favourable to development are well known: a realistic exchange rate; fiscal and price incentives for both exports and the production of domestic food supplies; availability of foreign exchange for transport and processing equipment, supplies and spare parts, and incentive consumer goods for rural populations; and confidence that a full range of marketing enterprises can continue to operate without abrupt changes in policy, arbitrary price controls, and harassment by local officials. A price policy that makes imports expensive benefits small producers, processors, and traders. It makes their output more valuable. The impact of such policies was evidenced by the recent expansion of agricultural production for export and to meet domestic food requirements in Ghana in the late 1980s. This was stimulated by high black market prices with only limited supplies on sale at official prices. Renewed availability of consumer goods has been a major factor in the revival of agricultural production in Tanzania.

Many African governments concerned to placate politically powerful urban consumers have kept food prices artificially low. This made the use of fertilizers to grow food crops unprofitable, hence the need for large subsidies that could not be maintained. A survey in Nigeria showed the unforeseen arrival of a shipload of low-cost imported rice to be the major hazard for the small domestic grower.

In addition to credit for the purchase of production inputs, finance is also needed in rural areas for marketing. Farmers and traders need credit to hold stocks to meet consumption needs and wait for favourable prices. This has long been recognized in India where public warehouses were established. They could issue certificates on produce in stores that were negotiable for credit at a bank. The produce could be taken out of store when the loan was repaid. This procedure is smooth enough with large standard lots that can be sold in store by description. For small farmers with small lots it is cumbersome. Most convenient for them is extension after harvest of a loan taken to pay for inputs. This need not present much difficulty where land titles can be offered in security. Where, however, allocation of the next year's loan for inputs is conditional on full repayment of the previous year's, it can disrupt the timely arrival of the next year's inputs. Group management of credit for holding stocks on the farm or in joint storage should accommodate differing views on its duration.

The scope for raising the value of small farmers' market output through practical advice and the application of research is great indeed. They need to plan their sales to the best advantage. Many small farmers lag in matching their output to market requirements. They tend to offer small lots surplus to family needs, irregular in variety, quality, and maturity. Such suppliers cannot be counted upon from one season to the next, so they do not get priority from buyers serving an established consumer clientele. Processing methods for farm and fishery products, the best form of packaging taking into account constraints on materials, use of market information to time sales to the best advantage – these are all issues for marketing extension.

Women

In most parts of the developing world, women play a major role in agriculture as workers, and in some parts as farm operators. Because the land, however, is in the name of the husband they have difficulty in obtaining credit, sometimes also in obtaining use of transport for marketing. Because of long-standing male predominance in mixed group activities, they have little say in decision-making by co-operatives. It is in protest against this male predominance that women's group savings and input purchase activities proliferated in Southern Africa, outside the co-operative system.

While in the long run it would be preferable to integrate women into the main line of rural institutional development, this may not be practicable for them until they have full control of their assets and the products of their work. It is in reaction against discrimination over access to land that women in coastal West Africa, the Caribbean, and the mountain regions in Latin America became so active in domestic marketing. Special programmes have been instituted to facilitate their operations – such as provision of child care and rest places at markets, and access to information and credit. The critical requirement for women in agriculture and its marketing is that they be relieved of discriminatory handicaps, at least in access to official services. In Gambia, women can obtain credit if they are registered leaseholders of land.

Government support

It is recognized that much less should be expected of government than in the past and more be done by local initiative. Even so, consistent government support can still be very helpful. A continuing handicap in many countries to a steady development of marketing and input supply systems in service to small farmers is dispersion of responsibility. Agriculture, commerce, rural development ministries, and local authorities may all be involved. Co-ordination can be difficult. Too often, decisions are taken without appreciation of their wider implications. A professional marketing support unit is needed at an appropriate place in the government structure to:

- assemble and analyse information and forecasts as a guide to policy on the development of private, co-operative, and state marketing enterprises and facilities as an integrated, internally competitive system, including ensuring the availability of credit, transport vehicles, and essential supplies;
- co-ordinate continuing support services covering market information, marketing extension, standardization of weights, measures and quality specifications, export quality controls, etc.;
- maintain a continuing focus on the use of labour-intensive procedures in marketing and processing, and implement measures to assist the smaller production and marketing enterprises.

EXPLOITATION

Small farmers are exploited when, for lack of information, because of indebtedness, or due to the absence of an alternative outlet, they are paid a lower price than necessary to cover the costs of the marketing enterprise concerned including a reasonable return for risk and management. Under the old private marketing system in Ethiopia many farmers were paid low prices because of high transport costs, wide margins, and some cheating over weights. Overall, however, grain markets there were said to be well integrated and prices reflected costs. Under the state system of the 1980s those farmers who had access to private buyers received much better prices than through the official channel. A review of marketing studies in Thailand noted the cost for farmers of breaking a credit tie with one merchant and moving to another, but observed that farmers were able to do it. Margins there were in line with costs except for some crops where transnationals had invested in new facilities and were using a semi-monopoly position to recoup their capital outlay (Siamwalla 1975).

The inference is that the small farmer is less likely to be 'exploited' where he has a choice of buyers, each with easy access to finance and transport, than when he can only sell to a parastatal – unless that parastatal is subsidized from non-agricultural resources. Access to a subsidized parastatal as a reserve outlet will, of course, strengthen his position. Arranging transport to one of its buying points would be a logical recourse for a group of small farmers dissatisfied with the other outlets open to them.

In a competitive environment, group action to bypass private traders will lead them to offer better terms. Unfortunately, such groups tend to break up later – unless there are special conditions in their favour as indicated in reference to co-operatives. Lowering the barriers to market entry is likely to be the most enduring way to promote competition and prevent exploitation. Under external pressure, President Marcos finally freed coconut oil exports in the Philippines from the monopoly assigned to the mills of his associate Cojuangco.

Unimpeded transport access is a vital condition. News of supplies awaiting buyers should then attract new enterprises. Provision of market information and making available transport vehicles and finance for purchasing and storage are the most direct measures. The entry of participants with a different profile of interests helps sharpen competition. This can be a buyer with his own retail outlet, one bypassing an intermediary market to serve directly a more distant consumer centre, one buying for processing, farmers who take up the marketing of their own and others' produce, and farmers' associations.

Limited sales space at assembly and central wholesale markets inhibits new competition, particularly if local authorities require all produce sold wholesale to pass through such markets. Pressure can be brought for the enlargement, relocation, supplementation, or bypassing of such markets. Reservation of space at assembly and central markets for irregular and seasonal operators, producers, and occasional traders, on payment of a small fee, eases their way into marketing operations.

EXPORTS AND THE RURAL POOR

There can be no doubt about the advantages of producing for export where the conditions are favourable. While earning the foreign exchange needed to pay for essential imports, the original settlers in the Gezireh became the aristocrats of agriculture in Sudan. Now their children pay immigrant workers to pick the long staple cotton. The returns from coffee production in Kenya in 1989 were about 30 cents per work hour once the trees were established, as against 6 cents from maize for local consumption. Farmers in Guatemala growing snow peas for the North American market grossed 14 times more per hectare in 1989 than from maize. Net of costs, their incomes have doubled. Income gains were highest on the small farms that provided their own labour and among landless households that found increased employment (Von Braun *et al.* 1989).

Production for export markets and for domestic food intake were set up as contrasting alternatives in socially oriented writings of the 1970s. There were situations where concentration on well-paid export crops left the local population short of food supplies. Food failed to arrive from elsewhere in response to the new market demand because of lags in marketing initiative and drought. This led tea growers in Rwanda, for instance, to neglect their tea and plant food crops between the rows. Most programmes which produce sugar, tea, tobacco, etc. for export processing now foresee this issue. Food crops are incorporated into the rotation. Otherwise, land is set aside for subsistence cropping.

More generally, FAO maintains that export cropping enhances food crop production because farmers learn from it the advantages of applying fertilizers and other inputs. This is well demonstrated in cotton production in countries like Burkina Faso. A food crop following cotton benefits from fertilizer residues left in the soil. Use of purchased inputs on food crops is also favoured where the proceeds of export crop sales are available to repay credit.

There is also the risk of fluctuating prices. International price stabilization agreements have been tried with limited results. National stabilization mechanisms based on reserve funds have been effective where the reserve funds could be protected against currency deterioration and the depredations of politicians, and where the prices announced for producers were realistic. Smallholders can obtain the stabilized price directly where located within reach of official buying agencies, otherwise indirectly via the stabilized market.

More difficult is price stabilization of highly perishable export crops such as green beans in Senegal. Here, management of seed allocations on the basis of market intelligence and diversion of surpluses to other outlets may be the best course.

There is a view that the small producer of export crops is less affected by price fluctuations than those with major financial commitments. This hinges on their access to family labour and to land at no continuing cost. Smallholders in Indonesia tapped their rubber trees when export prices were high and concentrated on food crops when they were low. The sponsors of smallholder tea developments in Kenya, however, insisted that growers face current world prices. As low-cost producers they have continued to prosper.

In the 1970s, international aid encouraged the governments of low-income countries to help the poorest of the poor. This placed a heavy burden on local administration. There could be revenue gains in the long run; in the mean time there were recurrent costs. The advice was well-intentioned, but derived from the conditions of wealthy countries able to afford special programmes for limited poverty sectors.

The approach to poverty alleviation outlined above is one designed to combine it with growth. Improvements in the poverty sectors strengthen the overall economy if they are inherently self-supporting. FAO's intention is to promote initiatives for progress through incentives to individuals. The main requirement of government is then to assure the necessary physical infrastructure, provide strategic support services, and maintain consistent policies for employment creation and initiative. The prices of imported capital equipment and of energy can be raised to promote the use of labour-intensive alternatives. Tax and other incentives can be provided for enterprise in putting people to work. Barriers to new entrants into marketing and processing can be removed, thereby increasing competition as well as opening up ways to additional income-raising activities.

9

FAO MARKETING
ASSISTANCE

Most countries, including the USSR, look to FAO for statistics and technical information on a world basis. It is also a convenient place to discuss problems of common interest. For these purposes a very much smaller and less costly organization would suffice. The justification for FAO's expansion in scale and range of operations over the decades to 1980 was the help it could give to countries needing technical assistance in its specialized fields: food, agriculture, fisheries, and forestry.

With evidence of underdevelopment and increasing poverty always before us, it is easy to query the work of FAO and ask what effect it has had. Its own Director General declares regularly that one-quarter of the world's population still goes to bed hungry. The conventional response of the experienced FAO staff member is 'without us it could have been worse.' He or she will then add, 'we gave them our advice, but it was not followed.' The reason it was not followed is that the countries were underdeveloped. Either they lacked the means to put the advice into practice, or those in authority did not consider it feasible or desirable. The basic problem – the political inability of many governments to put their full weight behind population control – has already been noted. Others show up in the following review of FAO assistance in the field of marketing.

ASSESSMENT OF MARKET OPPORTUNITIES
AND MARKETING ORGANIZATION

A request to provide a general marketing adviser was often the first step in seeking FAO assistance. Newly independent governments sought a fresh look at their economy; they were interested in policy leads that might not have been forthcoming from the former colonial power. Identification of export opportunities outside the metropolitan country and of scope for greater participation in marketing by the indigeneous people of the country was welcomed. Such an assignment could cover a range of agricultural products for the country as a whole, or concentrate on a defined area where incomes were low and a need for development was evident. Further help might then be sought in follow-up action. Here the focus would be sharpened – often calling for specialization in the marketing of particular sets of products, e.g. fruit and vegetables, livestock and meat, and grain. From the 1960s onward virtually all newly independent countries sought this kind of assistance from FAO.

Another common working base for a broad marketing adviser was as a member of a team working on argicultural development in a defined area. Illustrative would be one to plan farm operations on land to be irrigated by a dam on the Euphrates river. High value production was needed to justify the investment outlay on the dam, irrigation canals, access roads, and farm structures. Where could crops suited to the production conditions be sold profitably? What new marketing facilities, organization, and support services would be needed to assure this? The marketing specialist would be expected to identify export and domestic markets where rising demand would absorb additional supplies. He would have to assess the investments in transport, storage, and processing needed and the incentives required to attract competent marketing enterprises. Then he should estimate the prices likely to be paid to the farmers.

While confined to a single production base, such an assignment involved the balancing of various options. Professional consulting firms tended to present their findings in very clear terms, e.g. tomatoes for processing rotating with food grains and alfalfa for livestock, perhaps more definitively than was justified by the facts. The first man FAO put on to this job had experience and background welcomed by the government and the project manager. He collected a mass of information, but could not draw conclusions. In this he had to be helped by someone able to separate the big issues from the detail.

FRUIT AND VEGETABLE MARKETING

By the 1960s this had become a major line of FAO marketing assistance. Opportunities through climatic advantage to supply fruit and vegetables to consumers in areas and at times when they were not available from local sources could be a major stimulus for development. Exporting countries have benefited from:

- opportunities to earn foreign exchange by exporting high-value, out of season produce;
- the income-raising opportunities they offer to small farmers;
- the contribution to employment made by their labour-intensive production, handling, and sales requirements.

The expansion of fruit and vegetable exports from Africa and the Caribbean to northern Europe and North America has been one of the great successes of the last two decades. FAO's marketing advisers had a significant role in this. Lightweight, attractive, and protective packaging is now widely used in exports of fruits and vegetables from developing countries. An FAO adviser, George Holsten, designed a standard packing case for Kenya that could be produced at low cost, with individual exporters' own labels applied later.

Help was sought on a range of levels. A request from Greece came as the following telegram: 'Preliminary consideration fruit marketing problem indicates following specific detailed information required from agencies Western Germany, England. (a) Basic reaction distributors to prospects market entry new firms. (b) Comparative advantages canned and frozen products. (c) Specifications quality levels and packs each product, form. (d) Methods and costs introductory activities. (e) Minimum quantities assortments required for establishing, maintaining position. (f) Prospects private label versus contract packing including potentialities for combining the two. Consultant valuable this stage only if able supply details from intimate personal knowledge specific European firms.'

The plight of settlers in the East Ghor valley of Jordan attracted a UNSF allocation of $1.3 million for marketing assistance in 1964. They could not find satisfactory outlets for their produce. Packed and presented better it was expected that their tomatoes would fetch good prices from oil-rich consumers along the Gulf. Unfortunately, the government insisted on acquiring mechanical packing equipment and operating it directly. Experienced traders who had long-standing contracts with growers were left aside. The government went ahead with demonstration shipments of produce graded and packed with the new equipment; but it never attained a commercial level of operation. The government department concerned wanted to hold on to an activity that attracted external aid. When the FAO project came to an end it turned to the United States and other aid programmes. Twenty years later it was still the tomatoes of private wholesalers packed by hand into old apple boxes that were to be seen on sale at the import markets of the Gulf.

Another adverse political intervention was of the kind described in Hubert Creupelandt's 'Uncooperative melons' (Abbott 1986). An alien entrepreneur in Chad built up a profitable export of melons to Europe by air in winter. Work was provided for a number of people. It attracted the ambition of a local political figure, who had the enterprise taken over to be a co-operative, including his personal melon produce. However, when the melons were checked for quality those of the politician always went through. If space on the plane was limited, his melons were loaded first; those of others were left sitting on the tarmac. Exports dwindled, then ceased altogether.

A project much commended was that to assist the vegetable growers' co-operatives of the Kigezi District in Uganda, which had 4,000 members. This area is 400km from Kampala, the main market – too far for farmers to sell directly. Altitude favoured the growing of temperate climate vegetables. FAO provided as adviser a man who had a good knowledge of horticultural production, was a superb marketing organizer, and possessed a gift for getting his ideas accepted. To help adjust purchases to expected market demand he began with the allocation of seeds. He then had prices announced in advance for specific vegetables in seasons when it was known that there would be a short supply. This was done at the time of planting and backed up by field staff. The farmers reacted promptly. Good results were obtained with lettuce, potatoes, and carrots. Still there were periods of over-production when buying had to be restricted. The union management in Kigezi then worked out quotas for each primary society according to its membership and their production conditions. These were based on estimates sent by the sales staff in Kampala.

The society remained fragile, however, for lack of qualified members to run it. Finding and supervising someone who could manage their sales in Kampala was especially difficult. None of the members had the necessary marketing experience; few had even been to Kampala. This complex organization went down with the general disruption in the later years of Idi Amin. Marketing continued, however, with in-dependent traders serving the different groups of producers.

Many cities in the developing countries have been growing at a rate of 6 to 10 per cent per year. They doubled their population over eight to twelve years, and their fruit and vegetable supply system had to expand correspondingly. The heart of the supply system was usually a wholesale market. Here retailers located to serve consumers in the different parts of a city should be able to obtain supplies regularly, quickly, and at low cost. Growers who had access to transport, and traders who assembled produce from a number of growers, should be able to count on a reliable market outlet. The situation in many cities has been that either there was no such convenient central supply centre or that it was designed for a much smaller population and had become congested and unhygienic.

The FAO Marketing Group provided advisers to help design and locate new wholesale markets for many Latin American, Asian, and African cities. Plans, illustrations, and notes on experience were brought together in a briefing and advisory guide. One of the first major requests was to assist in the planning and construction of a new market for Sao Paolo. FAO was able to provide an Italian marketing specialist who could communicate easily with the local authorities. Selection of a site on a new ring road for convenience of access from producing areas was controversial at the time; the city had long since moved out beyond it. For a number of other large cities (Baghdad, Beirut, Teheran) plans for new wholesale markets were prepared but never implemented.

The time period between assessment of needs, preparation of plans, carrying out of feasibility studies and implementation of a wholesale market project can be long indeed. One of FAO's most successful market projects was that for Malta. Previously, trading took place at all hours in and around wholesalers' houses; they were scattered through the narrow streets in the old city of Valletta. Congestion was extreme, and most consignments had to be hand-carried in and out. Producers, wholesalers, and retailers were all agreed on the need for a new convenient site. One was available at no cost – an abandoned airfield with hangars that could be used for display and sales places, and ample tarmac for truck movement and parking. It was located *en route* from the main producing areas to the city.

FAO deployed a mature Dutch specialist with long experience of fruit and vegetable marketing. His assignment was for a year. Long before the year was out he had completed his assessment of the various interests involved, prepared detailed plans, and come out with a very favourable cost benefit ratio. He left Malta a frustrated man; nothing had been done. FAO received the usual bland note of appreciation from the government, then heard no more. Enquiries through informal channels had no effect. It was ten years later that FAO received an official communication from the government of Malta – the market would begin operations the following month. There would be an opening ceremony; could its adviser be present, whereupon he would receive a medal. Fortunately he was still alive and in shape to travel.

Establishing a market within his reach is one of the surest ways of helping the small farmer. During the late 1970s twenty model producer markets were set up in Brazil with FAO guidance. Their purpose was to link into the marketing system farmers located in areas with few outlets for their produce. These markets featured a raised area roofed against sun and rain, with closed accommodation for services. Local authorities were involved in their management; traders were contacted to ensure regular attendance. These markets proved potent factors for development. Access roads and communications were improved. Bank, rural credit, and extension offices were opened, and farm supply shops established. There was also a significant redirection of produce flow towards consumer centres of the interior. This avoided transport to an urban wholesale market and back again. The savings were reflected in prices to producers and consumers.

Upsetting for some external advisers was the handling of produce for domestic markets. They saw the palm stem crates used in Egypt as a cause of damage and loss for tomatoes. Yet these crates were still in general use there thirty years later. They were put together with local labour and materials. Cartons were available, but they cost $1.00 each and used up foreign exchange. Domestic consumers would accept cuts and marks from the local crates if the price was low.

LIVESTOCK AND MEAT MARKETING

In most developing countries meat has a high income elasticity of demand so that, when incomes rise, a relatively large share of extra income is devoted to meat compared with other foods. So programmes to expand livestock output and productivity have had high priority. They would lead to valuable exports or reduce dependence on imports. In either case they could raise herders' incomes and promote rural development.

There were many countries that sought FAO help in this area. The first were in Latin America. Traditionally, consumers in the cities were supplied from municipal abattoirs. Cattle were brought to these live, often travelling long distances without feed and under harsh conditions. FAO had an American specialist, Roger Burdette, who helped several countries develop systems whereby additional meat was brought under refrigeration from abattoirs located in production areas. It was then displayed in a refrigerated wholesale meat market where retailers could come to buy. His work on the rail transport of meat from an abattoir owned by livestock raisers at Osorno in Southern Chile attracted wide attention. A livestock marketing training meeting was held in Chile in 1955 to which people were brought from other parts of Latin America.

As a result, requests for assistance came in from other countries. A strategic role of the FAO adviser was to mobilize central government support against continuing resistance to change on the part of municipal authorities and established live animal wholesalers.

FAO's work on livestock in Africa south of the Sahara began with a survey in 1960. It stressed both the prospect of rapidly expanding consumer demand for meat in Africa, and the production potential that was largely untapped. The survey was followed up by a training meeting for people from all parts of Africa. This was held at N'Djamena in Chad, where a new abattoir had been set up with French assistance. Beef quarters from its cold rooms were transported by air to the West African coastal cities. The quantity marketed in this way reached 14,000 tons annually, before the drought of the 1970s cut down its live animal supply.

The availability of such 'technical solutions' led to a series of attempts to tap pockets of low-priced livestock in locations far away from consumption centres. Igor Mann in Kenya devised mobile equipment to convert low-quality range stock into an easily transportable meat powder. FAO found itself in a cautionary position. It urged careful assessment of the relative advantages of moving meat or live animals under African conditions. Many consumers there preferred freshly killed 'hot' meat, and were reluctant to accept a chilled product. Local slaughter was essential if this preference was to be met. Transport of meat called for major investments in refrigerated trucks, and stores along the route to the consumer centre, and at retail outlets so that the meat could be kept cool on display. The capital cost would be high, maintenance difficult and expensive and the risk of spoilage considerable. There are large savings in weight in transporting dressed carcasses instead of live animals. But in poor countries most of the fifth quarter (liver, stomach, etc.) goes for human consumption along with the meat. If it had to be left in the production area it might be worth very little. The costs and prices involved in an economic assessment are illustrated in Table 9.1. Moving animals on foot to the Ghana border and there transporting them by truck to Accra was $22 per head more profitable, than slaughtering in Mali and sending the carcass by air.

The abattoir for Maseru, Lesotho was a classic example of the interests conspiring to build 'cathedrals in the desert'. These interests comprised:

- national planners, looking for projects that would accelerate development;

Table 9.1 Comparative costs and margins, export of live cattle and carcasses from Gao, Mali to Accra, Ghana

Live animal	$	Carcass	$
Sale price Kumasi	163	Sale price Accra	189
Less Ghana import duty	23	Hide and byproducts	
	—	sold locally	3
Net sale value	140	Less Ghana import duty	49
		Net sale value	143
Costs			
Price paid to grazier	69		
Buying fee	3	Costs	
Vaccination	1	Prices paid to grazier	69
Export tax	3	Slaughter and cold storage	10
Trekking fee	8	Transport to airport	2
Transport in Ghana	11	Air freight	51
Transit duty (Upper Volta)	2	Weight loss	2
Ghana veterinary fee	1	Sales costs in Accra	1
Weight loss 5%	6	Total cost	135
Other costs in Ghana	6	Net return	8
Total costs	110		
Net return	30		

Source: Fenn 1977

- a bilateral aid agency eager to finance a plant for which a firm of its country had recognized expertise;
- a manufacturer of processing equipment and turnkey plants trying to secure as large a job as possible;
- a political leader looking for a cut on the contract.

The abattoir agreed on was far too large for the number of animals likely to be forthcoming, and under FAO's influence it was cut down by half.

During the 1960s especially, political ideology and popular feelings that a narrow set of people involved in livestock and meat marketing were over-charging for their services led a number of governments to favour co-operatives and state enterprises. FAO was asked to assist. Often it was clear that independent private enterprise would do better. Small labour-intensive enterprises had competitive advantages in the buying of livestock from scattered producers and in the retailing of meat to low-income customers. Flexible procedures and on-the-spot decision making were important in this trade. Advice to this effect could be rejected by the government concerned. FAO would then have to bury such recommendations in innocuous wording.

The development of dairy marketing in India has received continuing international support. The programme under which the animals supplying milk for the population of Bombay were moved out of the city and their milk marketed co-operatively attracted wide admiration. FAO built on to it a training programme for people from other countries. In contrast, very little was done in India to improve the marketing of livestock. This reflected sensitivities in the Indian populace toward slaughtering in general and that of cattle in particular. One morning in the mid 1960s I had a telephone call from Karl Olsen, the Nordic American who managed technical assistance funds in FAO at that time. 'There was a strictly confidential request from a highly placed officer in the Government of India. Could FAO help with a proposal to dispose of the large number of older animals that were consuming food, but giving no economic return?' A processing plant would be established in Goa. This had recently become part of India, but was still distinct as regards communications. The population was Catholic and had no aversion to the slaughter of cattle. Agents for the enterprise could buy up old cows in various parts of India, which would be driven or transported to Goa through an extended channel without publicity. Swift, the American meat-packing firm, would operate the plant – exporting meat extract, meat powder, bone meal, and hides. These products would bring in valuable foreign exchange. The major benefit, however, would be indirect – the land and other resources that would be freed for younger and more productive stock. The project was never taken up. Internal political opposition to the Government mounted, and with an election coming up such a risk could not be taken by a party which used the cow as its election symbol.

GRAIN MARKETING

From the outset, FAO's assistance in the marketing of grain was on a fairly sophisticated level. Most requests for help came from governments seeking to stabilize prices. 'Managing the market with 10 per cent of the crop' was the FAO marketing dictum. In developing countries, half the crop was commonly eaten in the villages where it was grown. The other half reached consumers through marketing channels. A shift of 5 to 10 per cent in yield at harvest could easily occur under rain-fed conditions. It would result in a much more than proportionate shift in prices. If the government could buy on average some 20 per cent of the quantity marketed it could keep prices within certain limits by holding and releasing stocks. Buying at a pre-announced price just after harvest would be an incentive for production, especially for the poorer growers under pressure to sell at that time when prices were lowest.

Already in the 1950s FAO had advisers in Colombia, Guatemala, Honduras, Jordan, Iraq, and Malaysia. Later it helped many of the African countries. For the Asian region it promoted an Association of Food Marketing Institutions to exchange experience and training.

Typically, a parastatal body was assigned capital to acquire storage and empowered to obtain bank finance under government guarantee to buy and hold stocks. The most economical approach was to use existing grain wholesalers as agents. This· avoided the overhead cost of maintaining staff to undertake purchases which were seasonally concentrated and could vary greatly according to the harvest. This did not ensure, however, that the smaller farmers received the guaranteed minimum price. So the stabilizing agency often had to set up its own buying stations in the rural areas.

To help meet the cost of these operations, FAO proposed that initial stocks be supplied by the World Food Programme and be replenished by it when drawn on to relieve famine. The stabilizing agency could also be assigned a monopoly of grain imports for higher income consumers, such as wheat or rice. Profits from these sales could be set against the cost of stabilizing the market for domestic food grains.

The operation of such a system called for considerable understanding on the part of the government. This was not always forthcoming. A grain price stabilization agency with the initials INA had been set up in Colombia: it did very little. An FAO adviser was requested, who assembled information and prepared an action plan, but nothing happened. The FAO man grew impatient and complained. One day he was called to the office of the president, at that time virtually a dictator. He was asked how the plan should be put into effect. He replied, 'by setting up purchasing depots with storage to buy at harvest time when prices were low.' 'Then build the first one here,' said the politician pointing to a map. It was a site 12 km from a rail line and not at all convenient. It belonged to the father of his mistress whom he wanted to please. Fortunately, this man went out of office not long afterwards, and INA went with him. A subsequent government set up the Agricultural Marketing Development Institute (IDEMA). It was to guarantee minimum prices to farmers and stabilize prices to consumers of wheat, maize, rice, beans, and potatoes. IDEMA is still handling the task for which it was created.

Yields of sorghum, the main cereal in Somalia, vary greatly. In 1967 the government obtained FAO/UNDP funding for the construction of storage for 20,000 tons, for seven heavy trucks, a mechanic to maintain them, and for specialized advisers and training fellowships. These resources might not seem much, but they gave the government confidence to go ahead. There were problems, however. We had appointed as FAO project manager an experienced man who had run the stabilization programme in Malaysia. One day he was back in Rome, dismissed because the minister responsible found him 'unco-operative'. The minister wanted to use the project vehicles for his election campaign! This had been refused. I was sent out to see the minister, a thin, wiry man with a mass of hair, who spoke excellent English. 'Mr Abbott,' he said, 'my interests are your interests. Your project depends on my support. So you should help me win in the election.' The next project manager proved *too* co-operative. Excellent cement was being imported free of duty by UNDP for use on the grain stores. The Minister thought it could also be used on his new house. Aware of what happened to his predecessor the new project manager complied, so he had to go.

Then came a revolution. Two years later, under a new government, the grain marketing project was re-started. This time there was no double-dealing. The earlier minister had gone to prison, and the project was operated by the army with full government backing. The agency established itself as an effective buying, storing, and selling organization. Later, I asked a man who had been a very successful UNDP representative in Somalia, 'What would you have said if the minister wanted to use your jeeps for his election campaign?' 'I would have suggested that he first try Oxfam,' was his tongue-in-cheek reply.

A similar FAO project to establish a grain price stabilization board in Botswana went smoothly from the start. It was helped to acquire a 6,000 ton reserve. For convenience in rotation this was tied in with the board's other operations. Here, it was the political leaders who were most helpful. The board was relieved of pressure in parliament to relax quality differentials on purchases from farmers by the minister of agriculture himself. Also a farmer, he declared that he would be the first to accept deductions where due.

In her implicit review of FAO's marketing assistance, Uma Lele of the World Bank judged grain market stabilization its most important contribution (Lele 1989). Some twenty countries were helped to establish a mechanism that could assure growers a minimum price. It could also be used to ease hardship to consumers due to poor harvests, and for eventual relief of famine. It constitutes the basis for Peter Bowbrick's claim that 'Working on marketing and price policy in the Third World you may make it possible to increase the cash income of millions of peasants by a quarter or a third, and so save thousands of lives. You may be able to prevent a famine by a shift in price policy or by an early warning system, as some of my colleagues have done' (Bowbrick 1988).

SEEDS AND FERTILIZERS

In Europe and North America farmers may be applying too many mineral fertilizers and chemical sprays for the good of the environment. In most developing countries they are applying too little. They are not obtaining the best yields they could from the land they work. Where rural people are poor because crop yields are low or their holding is small, one of the first suggestions is the application of suitable fertilizer. That they are not doing this is usually because fertilizers are not conveniently available or because they cost too much.

In the 1960s, higher yielding varieties of rice and wheat, and hybrid maize, opened the way to greatly increased output – provided the right amount of fertilizer was applied. So the distribution of improved seed and fertilizer became a vital development issue. The assembly and dissemination of practical information and experience was an important part of FAO's marketing assistance. The establishment of a working party on fertilizer marketing and credit with industry participation helped. It brought together annually specialists with a range of experience.

Country studies highlighted shifts in policy and the progress achieved. A fertilizer marketing guide adapted to Third World conditions was prepared (Wierer and Abbott 1978). A decade later, a World Bank official formerly in charge of fertilizer development in India said that for him this guide had been the most influential professional document he had read.

Provision of commercial inputs to the smaller growers of food crops often began as a government-sponsored demonstration. Moving to a low-cost commercial operation became a major development line. In parallel with it went the need for easy access to short-term credit. Where government hesitations over opening input supply to commercial initiative were overcome, as in India, Pakistan, and Bangladesh, fertilizer use expanded rapidly. However, in much of Africa during the 1980s it stagnated. Expensive distribution systems, transport uncertainties and the consequent need to finance stocks for long periods combined to put fertilizer out of the reach of most food crop farmers – unless heavily subsidized.

Fertilizer is bulky and cumbersome to store and handle. FAO stressed the advantages of direct delivery to the users wherever feasible, as against moving bags into and out of wholesaler and retailer storage. The goal of private retailers in India was to organize deliveries so that their clients could take their sacks directly from a truck on arrival in the village. This was carried a stage further in Zimbabwe. Distributors' field men arranged for groups of small farmers to receive balanced lots of phosphatic and nitrogenous fertilizers, seed and pesticide all loaded together on the same vehicle. This spared the farmer both retailers' margins and the trouble of obtaining these production inputs separately.

In the 1970s many developing countries wanted fertilizer manufacturing plants of their own to avoid the continuing outlay of foreign exchange on imports. FAO marketing specialists joined World Bank missions to assess the grounds for financing new plants. Often they concluded that an obligation to buy from a single national plant would raise the cost to the farmer. This did not deter commercial enterprise. One salesman was sufficiently adept and unscrupulous to convince simultaneously the governments of Kenya, Somalia, and Sudan that they should each buy a plant to make nitrogen fertilizer. All three governments were told that output in excess of their country's needs could be exported to the other two. Some second-hand manufacturing equipment was actually landed at Mombasa, but never put into use.

Transnational fires are still burning over the marketing of improved seeds. FAO's approach, for a long time, was to help individual developing countries build up their own seed supply and distribution systems. Mostly state enterprises, these progressed slowly. Much more attractive seed lines came on offer from the transnationals. How developing countries can use their know-how without becoming dependent on them is a continuing issue.

SUPPORT SERVICES FOR MARKETING

It was always understood that FAO could only give a lead in marketing improvement. Countries' own governments should be equipped to diagnose objectively the problems they faced and to take remedial action. Somewhere in the government structure there must be a unit with staff trained in marketing able to collect information systematically, analyse it, and formulate sound action recommendations.

Continuity was essential. Past experience could then be drawn upon when legislation or other intervention was envisaged, so that the consequences were foreseen.

A qualified marketing development unit was also needed to initiate and co-ordinate support services for marketing (market information, quality control, research, and extension) and to ensure that the necessary infrastructure, transport, and finance was available. A plan for such a marketing department based on FAO advisory experience is shown in figure 4.

Our greatest success in building such an institution was the Marketing Development Bureau of Tanzania (MDB). This was established in the Ministry of Agriculture with FAO/UNDP assistance in 1972. Its work programme included:

- market research and export promotion;
- training of marketing staff for government services;
- issuance of marketing intelligence bulletins;
- provision of advice on pricing policies;
- review of the operations of parastatal marketing bodies.

By 1976 the MDB was making an annual review of agricultural prices. It was the first attempt to present price proposals with a balanced analysis of relevant factors. Before there had been disjointed decisions on individual crops. An illustrative comment would be, 'It is probable that the 1982/83 announced price for sorghum will lead to purchases of red grain substantially in excess of domestic market requirements. The National Milling Corporation can expect to have storage capacity tied up by this product, and to incur heavy financial loss from deterioration in storage and loss-making exports, as was the case in 1978/79.'

The operations of the National Milling Corporation and various commodity marketing boards called for continuing attention, and the mobilization in depth of accounting and financial management expertise. In addition, the MDB was asked to monitor the food supply situation. International construction of storage, and provision of reserve stocks, resulted from its studies. The MDB became so strategic for the economy of Tanzania that the financing of external staff to supplement the personnel available nationally was covered by World Bank loans when UNDP funding came to an end.

Strong teams of experienced marketing specialists were fielded by FAO to start off and strengthen such institutions. Altogether, thirty countries were helped in this way. Marketing departments or sections became permanent elements in the structure of their governments. The

Intelligence and policy formation	Marketing development	Quality control, packaging, storage	Market information and extension

Assembles current supply, demand, price, and outlook data.

Provides prompt advice to the government on current issues and operations of government-sponsored marketing enterprises.

Needs marketing economists with practical experience, good judgment, and realistic perception.

Undertakes research and assembles information on marketing enterprises, channels, and facilities.

Advises the government, enterprises, and individuals on marketing conditions, methods, equipment, costs, and on investment projects.

Needs economic, marketing, and technical staff with a practical research and advisory orientation.

Recommends specifications for product quality standards, packaging, transport, storage for voluntary and compulsory use.

Needs marketing and technical staff, inspection personnel, access to laboratories.

Organizes daily and periodical market news services, prepares advisory material for use by extension services, promotes and supplements marketing training arrangements.

Needs staff able to present information in a convenient, easily understandable form.

Figure 4 Plan for a government marketing department

influence they exercised on policy varied with the quality of their personnel. They had difficulty in retaining high-calibre staff because the salaries offered were those standard for government services. Even so, the existence of the marketing department helped to maintain awareness of marketing issues and provided a convenient working base for subsequent short-term international advisers when needed.

The help provided by FAO government support teams varied with the needs of the country. In Zaire in the late 1980s priority went to ensuring that trucks going out to rural areas had the necessary spare parts and fuel. To provide a minimum of road maintenance, traders were charged with responsibility for the roads they used. WFP food to pay local labour was allocated through them according to the stretches covered.

That responsibilities for food and agricultural marketing were dispersed over a number of ministries and public authorities showed that many governments did not understand it. Nine such bodies were involved in Egypt in 1957 when FAO sent a specialist on fruit and vegetable marketing there. This was still the case thirty years later when an FAO/World Bank team preparing a project to finance new marketing facilities with the support of one unit found its proposals thwarted by jealousy in the others. Such positions were often so entrenched that the government concerned did little about them.

FAO recommended the establishment of a marketing council to bring the heads of such units together periodically, along with private sector representatives. It was rarely effective. In Iran FAO supported the establishment of a specific ministry for marketing. It was to be responsible for marketing policy and the state meat, sugar, tea, and other marketing companies. With such a portfolio, close relations with the ministries of agriculture, rural development and commerce were essential. The ministry failed in this and was abandoned. Still, staff had been trained. The senior FAO counterpart went on to become head of the National Fertilizer Distribution Company.

FAO has helped more than thirty countries to set up or improve public services to collect and disseminate information on market prices and prospects. This has always been a first line in marketing support. A guide on the operation of such services was issued in 1983 (Schubert, 1982). The second phase in such assistance was to check their performance some years later. Several were not continued very long after external financing ceased. Governments begrudged meeting the full cost from their own budgets. The absence of effective protest when the service closed must indicate failure to achieve its purpose.

Sometimes insufficient care was taken to ensure that the prices reported were meaningful to prospective users and reached them conveniently. Loose terms such as 'farm prices', 'official prices', etc. could be misleading. To be timely, price information had to go out on the radio. The best services used telephoned reporting by a reliable observer in the market. Some news services came into conflict with government policies to maintain maximum price ceilings or official purchasing programmes. A market news service started in Iran was suppressed when it revealed that actual market prices for many products were far above those set by the government and used in its cost of living index. In 1981 it was finally agreed in Kenya that prices obtained for grain from sales other than to buying agents of the Cereals Marketing Board could be reported. Previously these were kept under a veil, available only by word of mouth.

Hundreds of marketing extension staff were trained by the Indian Agricultural Marketing Advisers' Department in the 1960s for posting at local assembly markets. Few other governments followed this lead. The prevailing view was that marketing extension should be 'left to the co-operatives; they were the official development instrument in the rural areas'. Interest in professional marketing extension revived in the 1980s with general recognition that farmers must take much more initiative in marketing and that small private traders had an important role to play. FAO organized seminars to exchange experience, and IFAD financed marketing extension in its projects.

TRAINING IN MARKETING

When FAO began marketing improvement in the 1950s few people in development circles knew what it involved. It was just not part of their educational background. For most professional educators also marketing was a new subject. Only later was it incorporated into the teaching programmes of established universities and vocational training institutions. Meanwhile, those concerned with marketing improvement programmes were only too well aware that lack of competent personnel was a major factor in the slow pace of achievement.

FAO had three broad approaches in training. Each was intended to complement the others. They were:

1   Providing external specialists to work on current development activities alongside national counterparts. These would learn by doing, with the incentive of taking over eventually from the external specialist.

2   Sending selected personnel for specialized training in a more advanced country. Fellowships to finance this were often provided along with the funding of the international specialist.

3   Organizing group training programmes lasting from two or three days to a month or six weeks.

Requiring a government to nominate one or two counterparts for an external adviser was standard FAO procedure in the 1950s and 1960s. This ensured a working base and contacts for the adviser when he arrived. Through close association, the counterparts should acquire the confidence and ability to carry on when the adviser left. This arrangement could be very effective. An FAO man devised a quality control system for peanut exports from Libya; for the valuable edible peanut market they must be free of mould, etc. After some months his counterpart, called Ensor, went to study export quality control systems in other countries. On his return to Libya he took charge and the foreign expert left. Ensor then ran this service effectively for a number of years.

There were other cases, however, where the FAO adviser complained that: (a) his counterpart was too busy with day to day administration to give him much attention; (b) the counterpart nominated had no interest in marketing – Alex Thomson, working in Iran in the 1960s, said that his counterpart was a poet for whom his government post was a sinecure source of income; or (c) no counterpart was ever appointed.

During the 1960s, fifteen to twenty fellowships in marketing were awarded annually under the FAO/UNDP system. Not all of those who received them returned to take over the responsibility for which they were trained, but they added to the number of trained marketing people in the country. The timing sequence, starting with the arrival of the external adviser, was difficult to maintain in practice. Many fellowship holders tried to extend the period they were away, claiming they must complete a degree course. When they returned, the post had been filled by somebody else.

Often there was disagreement over the programme to be followed. The goal of many fellowship holders was to obtain a degree from a recognized foreign university. For them this was a solid step for promotion. Our view was that this was a goal to be approached through a government scholarship. Moreover, few of the degree courses available at that time were adapted to the needs of marketing in the developing countries. Instruction would be in terms of conditions so different (in the advantages of using technical equipment as against local labour for example) as to be quite misleading. Much more suitable, in our view, was an intensive short course in marketing followed by detailed observation of marketing policies, procedures, and operations in an intermediate-level country with conditions similar to those where the fellowship holder would eventually work. This came to be recognized in FAO projects by the 1970s.

From this time also, group study tours of short duration tended to replace individual fellowships. For the same total cost, more people were given access to the experience and conditions of other countries. Organizing such a tour was a considerable task. Marketing institutions and enterprises had to be ready to receive the group, and there were language gaps to be bridged. Once we had to prepare such a programme for fifty officers of the Turkish grain price stabilization organization. All of them, it was said, had equal entitlement to international training, but only a few spoke any language besides Turkish. They were divided into three teams. One, with a leader who spoke English, went to Canada; another, including a man with command of French, went to France, Belgium, and Switzerland; the third group, including several with a knowledge of German, went to Germany and the Netherlands. The language leaders would act as interpreters. The French and German-led groups fared satisfactorily. From Canada, however, came complaints from grain silo operators about an FAO sponsored group that lacked effective communication. But no one complained that they were not committed to their programme.

The third line of training was to bring together people responsible for marketing in a number of countries with a common language. Intensive courses, four to six weeks in duration, were organized for Latin America, the Near East, Africa, and Asia during the years 1955–65. The initial steps were to send round a man to ensure that the right people were nominated, and to prepare a dossier on their needs. Practice in the application of economic and marketing principles and analytical techniques was interspersed with demonstrations of product handling, grading, packing and quality control, and visits to producers, wholesale and retail marketing enterprises. Field surveys to assemble specific information on current issues were a valuable part of such programmes. There were countries where social traditions inhibited such contacts. The obligation to interview producers, wholesalers, and retailers of grain, eggs, vegetables, etc., consolidated understanding of their role in the marketing system and of their interdependence.

Satisfactory completion of the study programme entitled participants to a certificate carrying some prestige. As we were all relatively fresh from a university environment, we were inclined to be exacting. The local FAO representative would then be pressured into ensuring that all participants had a document of recognition.

The regional seminars were followed by seminars for a single country. These forewent exchange of experience, but they did promote a more open review of the problems and difficulties of the country concerned. When people from other countries were present, it was found that a veil was drawn over many domestic problems.

An essential task in preparation for such a seminar was to free the participants from their normal duties. This was most easily achieved when a prominent minister could be interested in the scheme. This could lead to a grand opening ceremony, obligatory attendance of influential government officials, and detailed reporting in the national press. Our goal of stimulating and extending awareness of marketing would then be achieved.

The regional training seminars attracted appreciative attention. They led some participants to propose that such training be institutionalized at convenient places. Proposals to this effect became resolutions at FAO regional conferences. Favoured locations were Bogota for Latin America, Beirut for the Near East, and Bangalore for Asia. Beirut was then a very attractive place and a growing educational centre. The local protagonists, Saab and then Cortas, each speaking English, French, and Arabic, saw the possibilities; but support from other countries was not forthcoming. Later, both men had high positions in FAO.

The Bogota proposal received wider backing. An FAO study of manpower requirements convinced government representatives at a Latin American regional meeting that many jobs in agricultural marketing would be opening up over the next ten years. They saw that the first ones to be trained, in what was in Latin America a new field, would be able to pick up the best positions. FAO had in Las Lorinez a potential manager with drive and vision. The Latin American Agricultural Marketing Institute (ILMA) was established in its own building with the backing of the Colombian Government, the Agricultural Bank, the Coffee Growers' Federation, and UNDP. A unique feature of the time was practical work in a marketing laboratory. Here trainees measured the moisture content and other quality features of product samples, and classified them according to grade specifications. In Latin America academic and social traditions ran counter to direct contact with agricultural produce and marketing operations.

ILMA trainees – 78 at postgraduate level, 165 by regular course programme, and 849 through short courses – came to occupy strategic positions in the market economy of Colombia and of neighbouring countries, and in advisory agencies. Ten years later they were still meeting regularly on the basis of the *esprit de corps* developed at the institute. There was also commercial research and consulting.

Crisis came with the completion of UNDP assistance. There was no charismatic local leader to take over from the FAO manager. The governments of Ecuador, Panama, and Venezuela, which had sent trainees, would make no financial commitment. The institute was continued on a much smaller scale at a private university.

In parallel with such projects FAO pressed continually for the inclusion of marketing studies in national agricultural and general teaching curricula. For universities established in the 'Oxbridge', Sorbonne, or older oriental academic traditions, marketing was a latecomer. Business training institutes were prevailingly urban in orientation.

FAO had already commissioned national assessments of the need for trained staff to meet prospective requirements in food and agricultural marketing over the coming decades. These attracted considerable attention. Few people were aware of the range of sub-disciplines involved and how little preparation there was for them in most countries. The summary assessment for Ecuador prepared by Mannarelli, a long-time FAO marketing adviser in Latin America, is shown in Table 9.2. The figures may seem high, but were considered realistic, allowing for wastage.

Technical meetings were held with university leaders in Asian and African countries. Marketing curricula adapted to their countries were developed. Recommendations on building these into existing teaching programmes were circulated. A major success was a tour of Asian universities by Michael Haines, who became Professor of Marketing at the University of Wales, Aberystwyth. His report appraised the instruction available and highlighted the lacunae. There were follow-up meetings in each country with UK and USA aid participation. I was taken to the one at Mymensingh, Bangladesh, in the FAO representative's car, but had to find my own way back to Dhaka. The Bangladesh chief of marketing had come in a jeep so I asked him for a ride. I didn't know that he was taking a goat home for the feast of the Id. For 150km I sat on the back seat with it, together with a handful of hay.

The outcome was substantial. Instructors from various universities were sent to the UK and USA for high-level refresher courses in preparation for more specialized course teaching. Marketing teaching was added to the curricula of provincial agricultural institutes in

Table 9.2 The need for trained agricultural marketing personnel in Equador, 1977–90

Programmes, projects, and services	Numbers to be trained		
	University level	Medium level	Specialized workers
Grain storage	58	330	240
Grain assembly centres (Co-operatives)	4	16	16
Marketing enterprises	24	48	48
Potato storage plants	10	20	20
Cold chain	14	36	36
Fruit and vegetable packing	8	16	32
Wholesale markets	30	60	60
Commission agents, brokers, auctioneers	30	20	—
Market and price information	20	30	—
Quality control	50	150	—
Inspection services	50	50	—
Supply statistics projections	20	20	—
Abattoirs and inspection	86	410	670
Livestock markets	30	60	95
Dairy industries, inspection	280	370	400
Meat industries, inspection	50	120	120
Instructors for marketing training	40	40	—
Consumer co-operatives and supermarkets	20	40	120
Total	824	1836	1857
Annual training requirements (1978–90)	63	141	143

Thailand. A marketing MA course was set up at Los Banos university in the Philippines. A full marketing department offering instruction up to a Master's degree was established at Bangalore in southern India.

With no systematic pre-service training available, the staff of major marketing enterprises and services in the developing countries were recruited on the basis of general education. Advisers on marketing improvement complained consistently of their technical and professional limitations, so FAO was asked to provide in-service marketing training. This was not at all easy. An agricultural business management professor, well known in the Philippines for his use of the case-study method, was defeated by the challenges which faced him in Kenya. His year there was wasted. In contrast, an Austrian, Werner Kley, was successful forthwith. His technique was to bring a group together and ask each person to describe his own job and then say how he could do it better. The others listened; then they were asked to make their own suggestions. Kley's services were in continuing demand.

Fred Scherer, FAO marketing adviser in Brazil, developed an effective training approach for workers in wholesale markets and employees of wholesalers and transporters. Fifteen per cent had no schooling, 49 per cent primary only, 33 per cent secondary or technical, and 3 per cent had been to a university. The topics which interested them most were:

- technical marketing and product knowledge;
- grading and packing of produce;
- business organization and administration;
- market information;
- market equipment and its operation.

They were prepared to participate in training sessions of one to four hours on days when they were not very busy. These took place at the wholesale market, using wholesalers' stalls for practical demonstrations. Slides and films on correct and incorrect ways of handling produce were presented, followed by discussions. Then groups of four or five performed all operations under the supervision of the instructor and experienced wholesalers. Requests for follow-up courses confirmed the usefulness of this approach.

SUSTAINABILITY

Can a regional development authority, a processing plant, or a marketing information and extension service be maintained after external financing ceases? Sustainability became a 'buzzword' in the 1980s when many governments of developing countries found their revenues sharply constricted. It had been an important consideration for FAO long before that. Almost automatically, FAO country representatives warned specialists starting a project not to recommend the establishment of new institutions – they should build on what already existed. Securing agreement that the government would support a project after aid funding ceased was of no avail if the government then had no money. So new activities and responsibilities must fit into programmes already budgeted, or generate their own income.

The Latin American Marketing Institute, mentioned above, was established on the campus of the National University of Colombia with the expectation that it would become part of the University. This was thwarted by academic and political disagreement. A contemporary commented, 'The university wanted the building, but not the subject.'

Some expensive lessons have been learnt on the establishment of development institutions and processing plants that could not be maintained or could never become self-supporting. But these were mainly an outcome of tied aid – engineering firms eager to set a plant in place and vanish with the profits, or higher minded groups wanting their own institution with which to 'identify'. FAO never had the money for such ventures.

Through conference papers, advisory materials, and training lectures, FAO has urged modification of parastatal and co-operative marketing and input distribution systems needing subsidy. It has campaigned for the involvement of local labour paid with WFP food in building rural infrastructure, and for the institution of maintenance procedures at the user level. It has stressed that maintaining a steady supply of spare parts and effective repair facilities can give much better returns to aid than furnishing successive fleets of new vehicles to specialized enterprises and projects. Public carriers constitute the most efficient use of transport vehicles in service to small-scale production and trading.

A general problem under conditions of inflation has been to keep cash charges for marketing support services and facilities in line with costs. Municipal and village authorities, in particular, have difficulty in keeping charges for market entry, use of abattoir facilities, etc. at a level adequate to maintain services and make repairs when needed.

It is worth considering in depth how to provide incentives for effective performance by local staff of intelligence, extension, and similar marketing support services after a project has ended. Initiated by, or under the supervision of, well-paid external advisers, operated for the duration of the project by local staff whose salaries have been topped up from project funds, such services can be very effective. Maintaining the same level when the staff remuneration falls back to the level of local salaries is difficult. Any extension man or woman who is capable of advising small farmers and traders on marketing will almost certainly feel that it is more profitable to work for a commercial enterprise or on their own account. An arrangement whereby they would participate in the benefits resulting from their work could be strategic.

EFFECTIVE HELP AT LOWER COST

Evaluation of technical assistance is usually unsatisfactory, and for this reason not often attempted. FAO abandoned 'post-mortems' in the 1960s as they found them unproductive. Only the formal aspects – Did an adviser complete an assignment without complaint by the supervisors or the government concerned? Were adequate counterpart and other support services provided? – are easily measurable. Impact on development may come years later when a counterpart finds it opportune to adopt an adviser's idea as his own and push for it, or when the government changes.

Most chiefs of technical support units in FAO have certainly been concerned that the assistance they offered be effective. They watched for indications of this in the reports sent back by their specialists, in statements from the country at meetings, and in personal contacts. They were also concerned that the funds provided should not be wasted. If they 'washed their hands of a project' it would be one that had been taken on against their recommendation. The following points flow from such observation and experience.

Periodically, FAO has found itself with projects that turned out to be aside from the strategic issues, or which attracted meagre government support. The internal view was that this must be expected. FAO was serving countries whose governments were also underdeveloped. They might not fully understand the advice offered or they may be nervous of the political implications. This was partly why they needed help, and one must work for small incremental advances that had good chances of being sustained. It is now recognized that external support for an improvement project should be elastic, to be suspended and taken up again by practical consideration, not strait-jacketed by administrative completion dates. It should also accommodate provision of materials and services beyond normal usage if a lack of these from local sources is prejudicing performance, and there is evidence otherwise of a serious commitment.

All the same, it can be asked whether FAO should have been more critical of the proposals coming to it for assistance. In fact, though its technical units were usually ready to take on any project that would add to their stature, FAO has not accepted all the requests it received. FAO representatives in the country concerned kept it away from some of the more impractical proposals and from ministries of poor standing. Later, we would hear that such a project had gone to one or another bilateral donor. At various times, FAO has sent out staff to help governments 'plan' their external assistance programmes. They tended to find themselves looking at one leg of an elephant, so to speak. Many Third World governments had become adept at playing off one aid agency against another and wanted to remain free to do so. One point is clear, however; while bilateral agencies have diplomatic reasons for not straining relations with a government over efficiency in aid use, a UN body *can* afford to do so – providing, of course, that there is no question of its director general needing that country's vote for successful re-election.

Could a greater effort be made to influence the thinking of higher echelon government personnel? Uma Lele thought FAO was too weak *vis à vis* the governments it was trying to help (Lele 1989). In the policy papers presented at conferences, criticism was watered down so that no government would be offended. Could more use be made of visits by eminent specialists to high-level policy makers? In some countries these policy makers change frequently, or suddenly; time spent on someone who will no longer be in office is lost. Others stay for ever and resist all attempts to influence them. I was given half an hour to warn President Kaunda of Zambia off parastatal marketing of perishables; it was to no avail. FAO technical managers usually concentrated on middle-level government staff in the expectation that they were relatively permanent and could become influential. However, the best ones tended to leave for more remunerative employment elsewhere. It is already claimed that there are many competent marketing personnel, in African countries especially, that are not being used in price and marketing policy functions. The best hope is that their increasing numbers will constitute an influential body of opinion.

Should FAO reduce sharply the proportion of its aid going to government institutions? Undoutedly, along with other aid agencies, it has been over-optimistic about governments' ability to keep running the programmes it initiated. Already by the late 1970s FAO had adopted an internal policy of building on to those already existing. The focus on privatization and self-sustainability in the later 1980s stemmed directly from the earlier overloading of governments with ongoing budget obligations.

The obligation to work through the government and with the counterpart agencies it nominated certainly induced a continuing bias towards state agency projects. In practice, the World Bank may have been the more tied to governments because of its need for an official loan guarantor. Of the agro-industrial project components financed by the World Bank from 1972 to 1983, 92 per cent were for parastatals and government ministries, 7 per cent for co-operatives, and only 1 per cent for private enterprise. But for massive World Bank and Scandinavian aid, it has been said, Tanzania would have given up the doctrinaire policies that impoverished its people much sooner.

Can aid be channelled directly to unofficial bodies to more lasting effect? FAO has been doing this to a modest extent with projects financed under the Freedom from Hunger Campaign, and Money and Medals. Increasing bilateral aid was channelled through agencies like Oxfam through the 1980s for work with individuals and private groups. In 1990 this was seen more coolly: there were biases and failings. In Zimbabwe, for example, a disproportionate amount of such aid, in the opinion of the ministry responsible, went to production co-operatives that showed poor prospects of using it effectively – because their chairmen were good at tapping gullible well-wishers. It is clear that the collaboration of a government is still needed on various aspects. Aid to private marketing enterprises in Tanzania, for example, would have been to no avail if the enterprises were still harassed by the police and restricted in the movement of produce.

Can technical assistance be less expensive? The teams of P 4 specialists sent to a country in the 1960s and 1970s for several years at a time have already been succeeded by a P 5 team leader supported by consultants who come for a month or two when needed. The leader provides continuity, the consultants specialized assistance. This is likely to give way to consultants who participate in initiatives for which the continuity is local. In UNDP jargon, 'the development process will be internalized' to build up national management capacity that has up to now been limited by inappropriate training and 'brain drain'. The consultant who joins a project for short periods at intervals can be very effective. He makes his contribution, then leaves counterparts to follow it up. A planned date for his return gives them a target for completing their work.

The Technical Co-operation between Developing Countries (TCDC) approach has great potential, as shown by the FAO project to assist food grain marketing organizations in Asia. Their staff benefited from the exchange of experience and training facilities between the two organizations. Most of the cost was met by the participating organizations, so the arrangements were economical. The return on the UNDP outlay for FAO support was high. Still, all the participating organizations were parastatals. The private enterprises which handled the balance of the food grain marketed in the countries concerned, perhaps 80 per cent, received no assistance at all. It should be feasible to provide comparable help to them through trade associations or chambers of commerce. Just as the European Association for Food Distributors organizes lectures and discussions for its members, programmes adapted to local conditions could be provided through national and district associations in developing countries. This would require somewhat more time and effort in preparation, but appears practicable.

Most of these approaches call for the services of people with substantial experience of developing countries, ample flexibility, and a tactful, persuasive style. FAO can contract them on a holding basis, with a minimum annual earning guaranteed, and the cost charged to the projects. Gaps can be filled by consulting firms adapting to these requirements.

Can the overhead costs of FAO technical assistance be reduced? Currently it charges 14 per cent on UNDP projects, and 13 per cent on those financed by other donors. For its own TCP projects there is only minor compensation for administrative and technical support. In contrast Oxfam, a voluntary organization, takes only 4 per cent. For commercial consulting firms, 50 per cent is the minimum. FAO raised its overhead charge from 8 per cent through 12 per cent on the grounds that its own regular programme should not subsidize the management of projects for others. If it has to operate in direct competition with consulting firms for UNDP-financed projects, cuts in its overhead charge can be foreseen. The most important consideration, however, will be performance. Ascertaining the availability of a suitable specialist, proposing the choice to a government for clearance, and then negotiating recruitment already takes long enough. There should not be further delay due to administrative inadequacy. Sid Galpin, completing a four month assignment in Gabon, wrote that he was 'being transferred to Senegal. Whilst in Gabon my salary had gone to my previous place of work. Now that I am leaving, I expect it will go to Gabon.'

10

THE MARKETING GROUP
IN FAO

In 1955 the FAO marketing group had just moved out of the Commodities Division. This division undertook commodity outlook reviews; it also serviced international study groups interested in stabilizing commodity prices on world markets. This was the main interest of the director, Gerda Blau. She had come to FAO from the International Wool Secretariat. The marketing group was to help individual countries strengthen their marketing. However, these two orientations could conflict. A commodity study group member – from the United States Department of Agriculture – once complained to FAO that its marketing group was helping a new exporter come on to the rice market.

The marketing group was moved to the Economic Analysis Division which had a major programme in price policies. Since marketing constitutes the mechanism whereby price policies can be implemented, this was a good fit. It was also very lucky for Marketing. The Director of Economic Analysis, Philip Barter, knew what it meant and had a favourable view of it; he gave the marketing group a free hand.

BUILDING A TEAM OF ADVISERS

It was Barter who allocated travel funds for a series of visits to developing countries to assess their marketing problems. The Marketing Guides were prepared and published with his support. It was the first guide on marketing problems and improvement measures, issued in 1958, that stimulated a flow of requests for help in marketing and the allocation of technical assistance funds to support it. At this time, there were three professional officers in marketing at headquarters and around fifteen on country assignments. John Tauber, Assistant Regional Representative for Latin America, proposed the appointment of regional marketing specialists in 1958. These were to focus on the problems of a particular region, and the incumbents should be fluent in its main languages. Tauber had noted the impact of the regional livestock marketing training meeting held in Santiago.

It was not easy to fill the regional posts, which called for a specific combination of professional and language qualifications. The incumbents they were not expected, however, to deal directly with all the issues presented to them. They channelled to headquarters requests for specialized or extended assistance and helped organize programmes with regional participation.

Outstanding among our regional officers was Chong Lee, a dynamic Korean who had studied at Harvard. His approach was to enable people to learn by working together and sharing their experience. He was a great exponent of the TCDC concept of technical co-operation between developing countries in advance of its official designation. He initiated a number of collaborative programmes between Asian countries – on marketing arrangements for very small-scale farmers; on rural market development; on the establishment and strengthening of fertilizer distribution networks; and on training in marketing at various levels. This approach, and his close contact with the people of Asia, led over the years to many constructive programmes. He had a great sense of occasion. At a technical meeting in Feldafing, Germany, on 'Marketing: A Dynamic Force For Development', he saw in the presence of German and British aid agency representatives an opportunity to gain their support for concrete projects. He abandoned the general paper he had prepared and presented specific proposals. The amounts of money contributed by the two bilateral aid agencies were small in total. But Lee's approach of maximizing the contribution of developing country institutions and people by offering a small incentive, and bringing them back together to report on what they had done, brought results far beyond those achieved by projects much more costly.

The heroes of FAO marketing assistance, however, were the field advisers. Headquarters officers faced with a question could consult colleagues, refer easily to professional literature, and look up reports on other projects. They worked from a comfortable office with competent secretarial assistance and telephones generally available. The person undertaking a field assignment had professional training and experience but would be called on to apply this under conditions often quite unfamiliar and inconvenient. The sense of isolation could be great, and local counterparts could be envious and critical rather than friendly. He could be kept waiting for appointments and information; discussion might go on around him in a language he did not understand; and interpretation could be inadequate. The demands on the field adviser's patience were immense.

Karl Olsen, in charge of FAO technical assistance funds within FAO for many years, looked favourably on Marketing, he said, because our field advisers were usually good. One of the most accomplished of our field marketing advisers was Sid Galpin. He was eminently qualified, having practical experience of commodity trading in the Indian subcontinent with Dreyfus. In Somalia his willingness to work night and day to protect damaged grain saved large quantities from further deterioration. In Libya he devised an improved machine for pitting dates that was greatly appreciated. He then helped to set up a canning plant at an oasis in the desert some 400 km south of the capital. It was on this that we were told he had contravened the Arab boycott of Israeli products – fruit cans marked 'Made in Israel' had been found at his plant. An Italian supplier had sent him samples, including cans prepared for sale to Israel with the outside printed to the buyer's design. It was not his fault, but he had to leave.

Sid's great achievement was the establishment of the Botswana Agricultural Marketing Board. He built this from nothing, preparing the enabling legislation, seeking finance, training staff, setting up purchasing depots, storage, and an accounting system. After four or five years the board was able to implement a minimum price incentive programme to producers and to stabilize supplies, in a country subject to recurrent drought. It supplied fertilizers, fence wire, and other essential materials to farmers at a much lower cost than before. His flexibility and willingness were great assets to the marketing group for more than twenty years. He took what seemed a practical approach without watching every administrative angle. People who pursued the latter course never achieved so much.

For newcomers to technical assistance, we provided an intensive practical and technical briefing. For men like Sid there was little to offer in technical support. Our best contribution was a lunch in Rome to which we invited colleagues with relevant interests and experience. Ideas flowed from the exchange of anecdotes. At the very least, all the marketing advisers had the feeling that they knew us and that we were interested in them individually, and in what they were doing.

For projects requiring a team of specialists to work together over a period of years, a manager was appointed. This post called for careful judgment, experience, and leadership potential. It was important that the research and recommendations of subject matter specialists be put forward to the government in a form that it would appreciate. There should be a short executive summary; unless the first pages provided valuable leads, a busy government official would not read any further. Action proposals must take into account the costs, ways of meeting them, and the implications for other ministries. When a number of advisers and consultants contributed to a project, it was up to the manager to ensure that their recommendations were co-ordinated within such a frame. Someone who had performed very well working alone, or in a team led by someone else, might not of course rise to this challenge. This came up in Kenya in the 1970s, when the project manager left suffering nervous breakdown. A successor was found with the facility required. Instead of sitting on a backlog of reports that never reached the government, he was soon feeding in short cabinet papers at the rate of one per week.

For the effective briefing of its field staff, the marketing group had to keep abreast not only of what FAO people had done in various countries, but also of the work of other agencies and of university research. This latter we came across by chance. An opportunity to maintain a select bibliography of marketing studies and advisory material came with a visit by Reavis Cox of the Wharton School of Commerce, Philadelphia. He saw that it would also be useful for university research, and arranged for initial financing by the Kellog Foundation; FAO maintained it subsequently. A chain of correspondents who reviewed marketing reports and publications in various languages and supplied notes on their contents, was developed. Bibliographical supplements based on their contributions were issued every two to three years. The marketing group also contributed entries to the overall computerized agricultural bibliography, AGRIS. While this could provide a comprehensive print-out of titles by keywords, the user still had to guess which ones were worth requesting and reading.

Marketing advisers had to adapt to a wide range of countries and conditions. The least popular through the years 1960–85 were Libya under Ghadafi, and Algeria. Both governments had chips on their shoulders about external assistance. A request by a minister for help in raising the efficiency of state marketing enterprises in Algeria was sabotaged by his staff. The FAO specialist provided was cold-shouldered. His telephone was cut off; finally his office furniture was removed. He was replaced by a Yugoslav in the hope that this nationality might be more acceptable, but he received the same treatment. When the Yugoslav received news of an appointment in Rome, he wept tears of relief.

Some countries were very slow in clearing advisers proposed to them. One man was eventually rejected by Malawi because he had not been married; this, they felt, made him a likely homosexual. In some other cases replies were delayed for months. It would then be discovered that the nationality was unwelcome, but the government did not wish to say so. This was a delicate area; the considerations behind government positions were never set down on paper. FAO staff had to learn not merely to avoid proposing Jews to Arab governments, but also that Indian and Chinese nationals might not be acceptable to some countries.

ESTABLISHING THE SUBJECT AREA

In the early 1960s, the marketing implications of some FAO activities were not recognized, or not seen as needing attention. As formulated initially, many FAO area development projects did not provide for work on marketing. Those focusing on dry areas around the Mediterranean basin were directed by macro-economists who believed that 'the plan was the thing'. Provide a suitable balance of production ingredients and physical infrastructure, and development would follow. A second group of projects centred on dam building and irrigation were put together by Andrew De Vajda, an engineer with formative experience in Russia. He was adept at convincing governments to go ahead where the physical conditions looked favourable. For the initiators of both these sets of projects, profitable disposal of the eventual increase in output as an incentive for continuing expanded production was taken for granted. The battle for the inclusion of a marketing component in such projects was won with the issuance of a paper showing how many such planned investments had failed, or been greatly delayed in achieving the results expected, for lack of attention to marketing. As Henry Ergas, then in charge of integrated development projects, said, 'We had made our point.' This paper, 'Marketing Issues in Agricultural Development Planning', went on to a wider professional readership in a book sponsored by the American Marketing Association (Abbott 1968).

Storage and processing are two marketing-related subjects for which technical responsibility in FAO has been dispersed. In storage, agricultural engineers had the lead; there was also a storage entomologist in the plant protection branch of plants division. The marketing group looked at the economics of storage as an element in a producer-to-consumer marketing channel. The three groups collaborated in the 1970 publication of a storage guide for tropical conditions (Hall 1970).

The World Food Conference of 1974 recommended that special attention be paid to the prevention of food losses from farm to consumer. For FAO this became an appealing base for voluntary donations. With the director general making the decision on which countries should be assisted, it was another source of patronage. The marketing group, which saw lack of incentive as a main cause of storage losses, had to keep quiet. A special office was established for the Food Loss Prevention Programme, making a fourth professional unit concerned with storage. Three of them, however, were in one division.

Responsibility for processing has remained scattered. When a governing conference in the 1960s approved a budget increase to strengthen FAO in food and agricultural processing, one professional post went to the agricultural industries group, one to the marketing group, and one to Commodities Division. Abattoirs and dairy plants were handled by Animal Production Division, with a focus on hygiene. The agricultural industries staff were all technologists.

The contribution of the marketing group was in assessing the viability of the processing enterprise. In the late 1960s Hans Mittendorf mobilized field officers in a range of countries to undertake a review of seventy canning, slaughtering, milling, and related plants set up in Africa that had all failed. These plants were adequately equipped technically. That they did not continue to operate was due to defects in management, in securing an adequate in-flow of suitable raw material, and in marketing their output. The study evidenced the high risk of disappointment if there was not more attention to marketing management from the outset (Mittendorf 1968). Thenceforward the marketing group had a part in processing development projects, particularly those for livestock, fruit, and vegetables. FAO 'bulletins' were issued, setting out processing methods for various products. Management of a processing enterprise for plant products, livestock, milk, and fish had to await an initiative by UNCTAD in 1985. Concerned with the export potential of processed products from the developing countries, it organized a debate on the issue raised by nutritionists and sociologists that processing diverted food resources away from local people. After review at a meeting of government specialists, the case-studies and conclusions were developed into an advisory and training text (Abbott 1988).

PRIVATE ENTERPRISE, CO-OPERATIVES AND PARASTATALS

When asked for a recommendation, the policy stance of the FAO marketing group through the years 1955–88 was for a mixed marketing structure – for private, co-operative, and state enterprises in competition. This was reflected in its marketing guides and in the professional content of its training. While the prevailing tenor of the 'planning era' of the 1960s, and the preferences of many political leaders, was for co-operatives and state enterprise, FAO marketing continued to quote Professor Bauer, the great protagonist of free enterprise and self-help, and point to the risks of monopoly and mismanagement. There was also Henri de Farcy, a Jesuit 'professor of distribution'. His books featured case-studies of development initiated by individuals relying on their own efforts. However, the developing countries could point to the success of agricultural co-operatives in Scandinavia and Japan, and to the wide use of marketing boards in the British Commonwealth. To say that conditions for their operation were quite different in many developing countries risked condemnation as racist.

Officially, as a United Nations organization responding to the declared needs of its member countries, FAO's technical assistance alignment was neutral. The marketing group was able at the same time to assist the government of Tanzania in the 1960s and 1970s to develop its state enterprise marketing system, and to help the government of Kenya strengthen a blend of private, co-operative, and parastatal enterprises. While the bulk of its field advisers were recruited from market-oriented countries, they included many of socialist persuasion. When the government of Zimbabwe sought an adviser on co-operatives with east European experience, FAO recruited a Hungarian.

On a number of occasions, advisers were asked by governments to assist with programmes and enterprises that professionally they considered ill-founded. Our view was that they should point to the risks and the costs, then do their best to make the project work if the government still wished to pursue it. Some were unable to restrain their criticism and had to leave.

In the early 1960s Philip Barter suggested that we prepare a 'guide' in marketing boards. He had a friend who had just been 'Africanized' out of the Nigerian marketing board system; we should take advantage of his experience. Hubert Creupelandt had worked with boards in French-speaking Africa. Together we produced 'Agricultural Marketing Boards: Their Establishment and Operation', Marketing Guide 5, 1966.

This became the definitive work on the subject. Its typology of marketing parastatals by degree of intervention into a free market structure was held valid through the critical climate of the 1980s. From this flowed a succession of requests for technical assistance ranging from the establishment of a multi-commodity export board for Togo to the management of food grain supply and price stabilization agencies in a wide range of African, Asian, and Latin American countries.

This did not mean that we were committed to parastatal marketing. In 1971 we brought together leaders of outstanding indigenous private marketing enterprises in the developing countries to delineate the conditions of their success (FAO 1972). A set of case-studies showing how private marketing entrepreneurs had promoted rural development was issued in 1982 (Harper and Kavura 1982). These reports circulated widely.

As an organization serving its member countries, FAO felt it should respect governments' political predilections and assist them in the policies they chose to pursue. The pressures for change came from elsewhere – from a people's decision in Sri Lanka where a private enterprise government was elected in 1977, or by an internal coup as in Guinea. Then came Deng's dramatic shift back from communism in China. For various African countries the commodity recession of 1981 was the turning-point. They had to go to the IMF and the World Bank for money to keep going. There, Thatcher–Reaganism had fostered a morale for making the liberalization of the more negative socialist positions a condition of financial assistance.

FAO leadership found it politically convenient to stand back from the strong line of structural adjustment. The first policy paper put forward by the marketing group, on the realignment of marketing and input supply services in Africa in the face of financial stringency, was edited down. In the following year it was issued unchanged, but a subsequent circular to governments through FAO country representatives offering advice on privatization was stopped.

FAO marketing had also to tread cautiously over co-operatives. While sceptical over the immediate scope in developing countries for an enterprise complex in both accounting and management, we had to respect the position of colleagues who were promoting them. The first influential warnings of the dangers of trusting to co-operative marketing systems as motors of development also came from outside – from Professors Hunter, who had worked in India and Africa, and Hyden, who was at the University of Dar es Salaam. Roy Spinks, FAO regional officer for the Far East, made similar points in a paper published in the

FAO Monthly Bulletin of Agricultural Economics and Statistics (Spinks 1970). This bulletin was in its time a valuable outlet for original articles by FAO professionals. Spinks attracted a rebuttal from the International Federation of Agricultural Producers revealing its ignorance of realities in the developing countries. Co-operative marketing and supply is needed in most countries to provide competition with private enterprise. How to keep co-operatives in business, without giving them a monopoly, is one of the great challenges in marketing.

RURAL DEVELOPMENT AND SOCIAL MARKETING

A 1969 internal reorganization of responsibilities in FAO saw the Economic Analysis Division focusing on development planning. At the time this was considered a prerequisite for development, and for the allocation of development aid. Marketing was transferred to the Rural Institutions Division, to be a 'service' together with credit and co-operatives, alongside agricultural education and extension, and land reform. In this company various areas of marketing went into distant focus. It was the cue for some other divisions – forestry, fisheries, and animals – to bring in marketing expertise as a complement to their production programmes. The International Trade Centre was set up in Geneva to supplement GATT/UNCTAD in helping the developing countries increase their export earnings. It provided direct advice on how and where to sell agricultural products internationally – an activity to which the FAO marketing group had never been able to apply a significant body of resources. When, as a result of a personal contact, the Government of Kenya requested FAO help in finding export markets for its meat, dairy, and horticultural products in the 1960s, the FAO officer responsible for technical assistance budgeting agreed only grudgingly. To him, such assistance was so specific as to be barely eligible for support from external public funds.

The work of the marketing group expanded in other directions. Input supply became very important. Many small farmers, particularly food crop growers, could not obtain fertilizer for lack of credit and convenient distribution systems. It was on fertilizer that the marketing group made its most telling use of marketing costs and margins. They were compared in the early 1980s for Asian and the African countries, and figures were contributed by universities and other qualified correspondents. Initially they were incomplete or grouped disparate elements together; eventually it became clear that various countries importing the same fertilizer were paying widely differing prices for it. For lack of information, or more private reasons, the buyers for official importing agencies were not seeking out the lowest cost suppliers. Help could then be provided to correct this. Many government-controlled fertilizer distribution systems had never recognized the cost of finance, yet it was often the largest single element in distribution. Rarely were discounts offered for early delivery to reduce this cost and that of the seasonal burden on transport systems of handling large quantities at one time.

At that time, fertilizers were sold by stated percentages of nutrient ingredients and in standard sacks. Comparisons of the initial price landed as an import or as sold from a domestic factory, and of subsequent costs and margins through to the farmer, were both legitimate and meaningful. In many other areas of marketing we felt obliged to point to the misleading conclusions that could come from crude comparisons of the spread between producer and consumer prices. The margin, for example, between the price paid in the USA by consumers of rice and that received by the grower was much larger than in most developing countries. This did not mean, however, that rice marketing in the USA was inefficient; the main element in its cost was the expenditure on packaging. Under the conditions prevalent in the USA this was essential if rice was to be sold. Ladled loose from a sack it would cost more in retail store labour and not be at all appealing to potential consumers. Awareness of the difficulty of making legitimate comparisons may have been too inhibiting an influence. It was Ahmed and Rustagi of the International Food Policy Research Institute, Washington that made the very significant comparison of food grain marketing margins presented in table 8.2.

Assistance to women was an issue for the marketing group long before it became a major FAO theme. In 1960 we proposed a training meeting for African countries on fruit and vegetable marketing. UNDP funds would be available if Africans thought it would be useful. This would have been an exchange of experience and information with specialists from each country. We sounded out the FAO regional representatives. Both the senior man (a Nigerian) and his assistant (a Ghanaian) were negative. Their reasons were the same; in their countries and others along the West African coast, fruit and vegetable marketing was dominated by women. Speaking only their local languages they would not be able to understand us, or each other. Moreover, in the opinion of our representatives, there was little that we could show them or that they needed to learn. Our representatives were also nervous. Market women were a major force in coastal West Africa. When Sekou Touré's government in Guinea tried to force them into a communist mould they raided the police station and tore up the files. In Ghana they were said to be behind a coup that brought down a government.

We went ahead with the training meeting. It was well attended, but not by the market women. We then tried to reach them through regional meetings set up specifically for women. One was held in English, one in French. This time the women had a voice; diffidence in making contacts on the part of predominantly male government personnel and educational limitations on the women's side had hindered communication. There had also been a feeling in some official quarters that the women's activities were temporary – primarily to provide funds to bring up and educate children – and that they were unlikely to found marketing enterprises that would continue and grow. On the other hand, their accumulated skills were certainly passed on by demonstration and emulation.

In follow-up, FAO implemented projects to improve the market facilities used by women, to provide child care services there, and credit for working capital and equipment. Perhaps our most significant contribution, however, was our continuing publicization of the role in marketing already attained by women in many countries and the need for a corresponding attention to their requirements, including equal rights with men in access to bank loans and property titles. In coastal West Africa, women handle 60 to 90 per cent of domestic farm produce from point of origin to consumption. They supply inputs on credit to farmers who contract their output to them. Women have a similar role in some Caribbean countries, and in the Indian-populated mountain regions of Latin America.

In 1964 the marketing of low-cost protein foods became of significant interest. Marketing was working in collaboration with the Nutrition division. Our goal was to devise distribution channels and promotion strategies that would bring these foods to low income consumers and vulnerable groups on a self-sustaining basis. Mostly they were cereal flours with fish or soy flour added. There were palatability problems, and the packaging needed to maintain stability made them too expensive. Progress with trial projects was reported to a Protein Advisory Group in New York, supported by the UN agencies concerned. A guide on the introduction and distribution of such foods was prepared by Bo Wickstrom, marketing professor at Gothenburg (Wickstrom 1974). Eventually an FAO/WHO expert committee concluded that rice and wheat contained enough protein to maintain people in reasonable health and these programmes disappeared.

For a year or two fish and soy proteins were succeeded in interest by protein products grown on hydrocarbons. OPEC pricing of petroleum products from 1973 put an end to this. Since then, socially oriented marketing has concentrated on two-price systems for food grains. The goal was to devise and help countries implement marketing systems whereby an adequately nutritious form or quality could be made available to low income consumers at prices they could afford, while maintaining a market remunerative to producers for the bulk of their sales. Much imagination has gone into 'self-targeting' by product variety, appearance, or taste and by the timing and location of sales to minimize leakage. Such programmes have proved practicable in Asia, but much less so in Africa.

By 1973 the Rural Institutions Division had been renamed 'Human Resources'. The 'Greens' of northern Europe wanted a 'human' focus. Marketing was moved out as essentially commercial. Together with agricultural banking and credit it went to Agricultural Services alongside processing, storage, and the prevention of post-harvest losses. The World Food Conference of 1974 was called over a situation of food scarcity. Marketing had no role on its agenda. Food security and post-harvest loss prevention programmes were established as targets for voluntary government support. They called for substantial marketing expertise, but were kept separate from the marketing group. This was the era of food systems – the various activities involved in bringing food to the consumer. While the proponents of this approach saw clearly that these systems had to be integrated into an effective marketing channel by a continuing business initiative, this linkage was played down in FAO. There was an external voice for marketing at the World Food Conference, in the form of the representative of the Association of Self-Service Food Stores in the USA, but he had a minimal hearing.

The marketing group in the Agricultural Services Division continued to be the central point for advice and policy in FAO, but marketing work also went on in a number of other places. In 1976 the group had a headquarters professional staff of eight, four regional marketing officers, and sixty-three specialists on country assignments. In 1989, with the US and other government contributions lagging, and posts frozen, the professional staff at headquarters of the marketing group was reduced to three. There were still twenty-four field advisers, and over forty short-term consultants. An assignment of a few weeks financed from the technical co-operation and similar budgets can involve as much work for the support unit as a long-term adviser. Forty per cent of the correspondence going through the AGS director's office related to marketing.

APPRAISING PERFORMANCE

How effective was the marketing group? Its staff were selected carefully for professional competence and practical orientation. They were able to make contributions to international journals that kept them in the forefront of their profession. When the senior officers left FAO they were in wide demand from other organizations. There were only two or three 'passengers' over a period of thirty years – specifically those who were recruited to comply with the requirements of geographical representation; they moved out eventually to other posts.

Leadership was consistent and positive over a substantial period of time. The advisory and training materials published and distributed were professionally respected. Roger Savary of the International Federation of Agricultural Producers, for example, liked them because they linked economic principles with practical action. They were also widely read.

The marketing guides were best sellers in FAO, reprinted many times to meet commercial demand. In addition to the English, French, and Spanish editions put out by FAO, they have been translated into and published in Arabic, Japanese, Turkish, and other languages. In the middle of the Afghan civil war permission was requested by a professor in Kabul to translate Guide No. 1 into Dari.

There was a healthy respect for the value of money. Continuity in management enabled the marketing group to predict that under the FAO annual budgeting system, additional funds would be available that had to be spent before the end of the calendar year. It would have project proposals ready for this that would build on what had been done before. This earned for the group the reputation in budget management offices that it consistently got more done than most with the funds allocated to it.

Many FAO field advisers were highly dedicated. Sid Galpin was so enthusiastic about the development possibilities of an oil-seed crushing plant in Botswana that he was prepared to put his own capital into it. To do this, however, he would have had to leave the organization. Aziz El Sherbini received a gold medal from the King of Jordan for his work on the Amman central market. Two governments nominated Fred Scherer for a B.R. Sen prize.

Some comparisons can be made with the marketing assistance available through bilateral aid. The main alternative was US AID, and its predecessors. Initially, we at FAO thought we had a much better grasp of conditions in the developing countries and of the improvements that they could maintain than US AID. We had the Indian model to draw on. The government there had carried out detailed surveys of the marketing of various commodities and identified the problems. It had set up a marketing adviser's department in the Ministry of Food and Agriculture. This was to provide a lead in formulating and implementing legislation to regulate rural assembly markets, the provision of storage as a service for others, and the use of quality grades for export consignments. Behind this lay a clear concept of a government providing support services for marketing by private enterprise. It was framed for conditions where farms were small, manpower abundant, capital scarce and government resources limited.

Many American advisers were unaware of this experience. They looked back to their own 'family farm', 'do-it-yourself' traditions. But this experience was drawn from a background of 100 hectare farms with easy access to mechanical equipment – not very relevant to most developing countries. The bulk of our advisers already had years of experience in Africa and Asia. US AID also suffered commodity taboos; we could help the government of Turkey improve export quality in cotton and tobacco when this was forbidden to US/AID by directive of Congress.

An advantage that the Americans had over FAO was that they could often mobilize cash to back up their advice. Funds from the sale in a country of surplus American commodities (the PL 480 programme) could be used to build a bridge, or top up the salaries of local counterparts to give them more incentive. American aid became more professional when universities were encouraged to concentrate on specific aspects of development. Michigan State University became the lead source of know-how in marketing, particularly in Latin America. University teams fielded under aid contracts provided opportunities for graduate students to gain experience, or material, for a thesis; they also nurtured respected leaders such as Harold Riley, Charles Slater, and Kelly Harrison.

Many young people became available who had been volunteers in the Peace Corps. They combined practical experience of developing countries, specialized university training, and a deep personal interest in their work.

In the 1980s provision of marketing support services to US AID headquarters in Washington was opened to competitive bidding by consulting firms. They could provide outstanding individuals, such as Edgar Ariza-Nino, but with high overhead costs. Their briefing of new people going out to field posts may be constrained by the need to charge all support time to specific projects. Thus when the firm on the support contract in 1989 was asked to brief the Lesotho Minister of Agriculture for a few hours during a visit to Washington it sought a fee of $3,000.

None of the other aid programmes has provided marketing in depth. That of the UK has been strongest on the technological side, with the professional backing of the Tropical Products Research and Development Institute. An adviser in the 1960s who encouraged a number of small countries to set up marketing boards for perishables may have done them a disservice. Scandinavian aid has been biased by the long-standing commitment of its guiding committees toward the promotion of co-operatives. The very substantial aid provided by the German Ministry of Co-operation, directly and through the German Development Foundation, has been open to the private enterprise alternative. Indeed it financed the meeting of private marketing entrepreneurs that FAO organized in 1970, a time when it would have been difficult to obtain funds for this from another source. The telephone market news service for vegetable growers in Java, initiated by B. Schubert, is a model of its kind – it is hoped it can be maintained when external aid ceases.

The professional group in FAO with the primary responsibility for marketing improvement has been successively a section, a branch, and a service, together with credit. It has remained quite small with a regular programme budget of about $600,000 per year – around 1 per cent of the total budget for economic and technical activities. Of the UNDP funds allocated to FAO in 1976 for technical assistance, some $3.5 million went to marketing – less than 2 per cent of the total. The key role of marketing in FAO's mandate notwithstanding, the unit responsible never rose above half a service. It remained on a par with planning, price policies, farm management, processing, storage and agricultural engineering as a group within a division. It has been moved three times in successive internal reorganizations, has been located at various times in four different divisions, but has never attained divisional status on its own. The appointment as Deputy Director General of FAO in 1963 of a man who had been head of the large Agricultural Marketing Service in the United States Department of Agriculture was first thought favourable for marketing. However, no significant change in its status ensued. FAO ended up in the 1980s with a small but professionally influential central marketing unit in AGS plus marketing elements incorporated into the programmes of various other divisions.

The decades from 1965 saw the attainment of independence of a large number of developing countries, and rows of new faces at the FAO governing conferences. They were concerned about the prices of food faced by their growing urban populations and the disturbing effects of sharp changes in prices to producers and consumers. They sought help in these areas, but their interest in FAO's structure seemed to go no further than seeking senior posts for themselves as a recognition of membership. One delegate from a new member country had the straightforward brief, 'On any spending of money say yes; on anything else keep your mouth shut.'

A clear stance of the FAO governing body in the 1970s was for a social orientation in its programmes. Delegations from the Scandinavian countries and the Netherlands took the lead in pushing rural development, in channelling assistance to the poorest of the poor. This was an era when the term 'agribusiness' could not be used in FAO because it smacked of 'big business', and when a Scandinavian government allocated funds for a rural development project, stating specifically that it should not include marketing because they did not want more consumerism. The programmes presented for marketing had to feature services to small-scale farmers. Much more money would have been allocated, it was once said by the FAO Chief of Programme and Budget who managed the governing conferences, if our main thrust had been in support of women.

Marketing was promoted by individual delegates to the FAO governing conferences, and in its programme committee. But the chairman of this committee in the 1970s was a Yugoslav; he declared that he did not understand marketing. The committee never made a move to raise its status. Government representations at the FAO conference came primarily from departments or ministries of agriculture. Forestry and fishing departments were also represented for countries where these were important. Food was covered by nutritionists. Marketing had little voice because in most FAO member countries it had no professional constituency seeking representation.

11

CAN FAO BE YOUNG
AGAIN?

The priorities set for marketing in FAO's programme of work, and the resources available to it, underwent substantial changes between 1945 and 1990. This has been true also of the organization as a whole.

SHIFTS IN WORK EMPHASIS 1945–90

Initially FAO collected and disseminated information, promoted research, and reviewed policies. A great event in its history was the release of its annual publication, 'The State of Food and Agriculture'. Its founders, however, had a grander design. They foresaw an organization that would manage the world's agricultural output to improve the well-being of its people. The first director general, Sir John Boyd Orr, had been one of the founding group. His ambition was to head a World Food Board that would:

- stabilize prices of agricultural commodities on world markets;
- establish world food reserves against periodic crop failures;
- finance agricultural surplus disposal on special terms to countries where need was urgent.

He also wanted to arrange loans to provide developing countries with agricultural equipment.

In 1946, this was considered unacceptable or premature by the developed countries which then constituted the main FAO membership. Since then, international price stabilization has been tried with moderate success, or less, for a number of commodities. Food reserve and surplus disposal programmes have been implemented over many years by the government of the USA. They were incidental to its internal farm price support operations, but served the needs of developing countries at the USA's discretion. The World Food Programme was set

190

up at the instance of other developed countries with surpluses to move; they were concerned to bring disposals by the USA into a joint decision-making mechanism. WFP has gone some way to meet the second two objectives of the original World Food Board, though much larger food supplies go to developing countries on concessional terms outside it. Its first Executive Director, A.H. Boerma, had the World Food Board concept at heart.

FAO had a role in the establishment of WFP, and provided administrative services for it until recently. The WFP executive director is appointed jointly by the DG of FAO and the secretary general of the UN after consultation with the Committee on Food Aid Policies and Programmes, to which both organizations nominate members. A portion of WFP food resources is reserved for use by FAO's director general in meeting food emergencies. WFP, however, has become increasingly independent.

The World Bank and IFAD are quite separate, but through them Boyd Orr's fourth great ambition, the provision of agricultural production inputs and equipment, has been realized. FAO's investment centre helps to identify and prepare projects for these financing agencies.

In the 1960s and 1970s FAO was providing technical assistance on a grand scale directly. It began with the transfer of knowledge and skills through the provision of expatriate advisers and specialized training. With the advent of the UN Special Fund, teams were mobilized to bring together various disciplines and achieve development breakthroughs that required concerted effort.

UNDP had financed some 12,000 FAO technical assistance projects by 1980. Since then, UNDP finance has declined. FAO has continued to provide technical assistance with some 40 per cent of it funded by bilateral and other donors using FAO as a professional executing agency, and by FAO's own Technical Co-operation Programme.

The 1970s saw a special effort by FAO and the World Bank to help the poorest segments of rural society. This was also a main objective of IFAD. A healthy preference for 'grass roots' projects continued through the 1980s, along with recognition that they must become self-sustaining. Aid agencies could not expect governments, themselves poverty-stricken, to maintain expensive projects after external financing ceased.

The 1964 agreement with the World Bank whereby FAO would help prepare projects for its financing was strongly positive. FAO's technical competence would help to guide World Bank finance. While the Bank could still take up projects in agriculture without reference to FAO, this arrangement put an end to the joke that FAO only provided advice and had not the means to help it into effect. In the 1980s, one-third of World Bank lending went to agriculture, a remarkable increase from the period preceding FAO partnership. In the later 1980s the formulation of new projects in agriculture became less important as more Bank money went to meeting current import requirements, with structural adjustment a condition.

Great capacity to expand food production has been evidenced. In the USA and western Europe, output has been restrained to avoid accumulation of surpluses. Increases in food output have to match fairly closely the rise in population; excess production can result in farmers going bankrupt and crops having to be diverted to keep prices at a remunerative level. At the same time there are countries, particularly in Africa, where per caput production has declined and the total number of people without enough food is increasing. The obstacle is poverty – people are too poor to buy all the food they need; governments are too poor to support social security programmes that would help their people in need. Average income per person in the developing countries has risen by 50 per cent since 1950, but it is distributed unequally between them and within them.

The executive director of the World Food Council summed up the situation to its conference in 1978. The developing countries blamed their continuing poverty sectors on inadequate support from the developed countries and the international agencies. For the developed countries, the fundamental cause was insufficient commitment on the part of the governments of the developing countries. The international agencies said their priorities were determined by their member countries; they were doing what they could with the resources available to them. No one of these three participating groups saw itself as at fault. This is still the case.

WHAT TO DO NEXT?

Conditions are now favourable for more selective and conditional approaches to international aid. It could be reserved for projects and activities that are specifically constructive in the recipient country. This would also help counter the view based on past experience that much aid is both wasteful and counter-productive. The reform of international aid is now both practicable and necessary:

1   Relaxation of capitalist–communist political rivalry has reduced

the motivation for financial support to client states on political grounds. Corrupt and reactionary regimes need no longer be supported in the interests of 'stability'.

2   That aid can be an effective lever for substantial changes in the recipient governments' domestic policies has been demonstrated by the structural adjustment loans of IMF and the World Bank. A few countries resisted, but over a score felt constrained to observe the conditions set. The same kind of leverage could be exerted for population control campaigns where needed. It can also be used to induce the deployment of resources away from armaments and private transfers to foreign accounts, and toward the building of a self-sustaining base and better services for the rural poor.

3   The plethora of UN agencies with their repetitive ineffective resolutions and duplicating post establishments has reached the point where public opinion would endorse financial pressure for reform. Many developed countries have tired of UN agencies set up by a majority vote at a conference where they had little say, yet are still obliged to provide the bulk of the finance. They see too many such agencies, too much politics, too much bureaucracy, too many meetings, too much paper. When a UN planned paper plant was set up in Central America the question arose, where would it obtain its raw material? No problem! It would recycle UN documents until its tree plantings matured. From 1978 to 1979 FAO issued 483 million pages of documents in five languages, and held 261 meetings. Posts are assured for developing country nationals, though their countries are short of skilled people. By 1980, 37 per cent of FAO's staff came from developing countries.

The implication for development aid is that it should be given only where it will be clearly positive in effect and that it be economized *vis à vis* reactionary regimes and superfluous UN establishments. This is a policy line for the aid givers in bulk – the International Monetary Fund and the World Bank, Japan, the EEC, USA, and USSR. A constructive role for the UN would be to promote understanding of the longer-run advantages of such an approach, and to work for co-operative policies in recalcitrant countries.

FAO's mission should be to keep alleviation of rural poverty to the forefront. To this end it should help poor countries to:

• keep domestic food production rising in step with the size of their populations, so far as this can be done at an acceptable cost;

- export agricultural products to earn the foreign exchange needed for essential equipment, etc., where these offer the best potential;
- establish policies and arrangements to raise the incomes and living conditions of the rural poor.

Achievement of these goals will depend greatly on:

- systematic expansion and improvement of irrigated agriculture to reduce the impact of erratic rains;
- steady improvement of crop and livestock production technology and management;
- better marketing of food and agricultural products in its broadest sense.

So the criterion for reform in FAO would be the paring down of its unproductive 'representational' structures and the redeployment of its resources to secure a clear focus on these goals.

FAO SHOULD BE PROFESSIONAL

FAO is by no means the only body that sees a role for itself in these areas. There are a dozen other UN-related agencies which do so also. In Africa specifically there are fifty intergovernment organizations, and 100 bilateral and non-governmental programmes. So FAO should specialize. While remaining sensitive to developing countries' needs and aspirations, it can let them channel their political pressure through the UN, its regional commissions, and UNCTAD. FAO can concentrate on being the world's best source of knowledge and assistance in its field.

Some of the bilateral and voluntary aid agencies interested in food and agriculture have found a development niche. Many others pick up and implement proposals that FAO could design and put into effect better and at lower cost. This is done in part to maintain the identity of the aid giver. It is also done because of FAO's low standing with their directing committees. With new direction, a 'purer' image, offering prompt accountability of the projects financed, FAO would attract much more of this 'business'.

Through the last decade FAO has struggled to maintain its technical assistance against a backdrop of declining financial support and adverse public relations. Should it continue to hold out for the principle of every member country an equal vote on all issues, no matter how small its population and how little its financial contribution? Should the vote of Grenada, for example, with 70,000 people and food from WFP to ease its foreign exchange deficit, carry the same weight as that of India with over 700 million people, or Japan with 200 million and a contribution of 20 per cent of the FAO budget?

Is it worth continuing with conferences that go on for weeks, take up the time of thousands of people and in the end rubber-stamp what is put before them because they are too unwieldy to do otherwise? Should FAO go to great lengths to maintain the 'support' of its member countries when what is meant is to secure their votes for a director general who wants to be re-elected for another term?

The traditional bi-annual regional conferences are also an expensive luxury. They oblige the organization to formulate a programme and mobilize interest, and the governments of the region to prepare formal speeches and to send representatives. Instead, meetings on specific issues of interest to a group of countries could be called as the need is discerned, to be attended by appropriate FAO officers and people from the countries concerned. The director general could make the opening of such a meeting a cue also to visit some of the countries. The normal procedure would be to hold such meetings in the language generally used for international communication, e.g. Spanish in southern and central America, English in the Caribbean. Issuance of conference papers in translation and simultaneous interpretation would be provided where there was an evident need, e.g. to facilitate inter-communication of African countries on a matter of common concern, not just to accommodate one or two small countries. An FAO staff member commanding their language could visit them, report on the meeting, and bring back their views.

Specific measures to raise morale and efficiency in FAO and commend it to major sources of development finance include the following: eligibility of a director general for one term only; the grouping together of small countries for votes on strategic issues; the abolition of elaborate representational structures; restoration of the prestige of the technical departments; sharpening of staffing policies for morale and efficiency; and making field projects fully accountable.

Eligibility of a director general for one term only

This can be of five or six years, with no re-election. It has been said of the last two directors general that one spent too much time travelling to ingratiate himself with member governments whilst the other distorted the FAO's Technical Co-operation Programmes and its public image to this end.

A convention that directors general should be elected alternately from developing and developed countries would help to maintain developed country support, give the developing countries the feeling that the organization was theirs also, and bring to it refreshingly different experience and viewpoints.

Group together small countries for votes on strategic issues

These would include election of the director general, changes in the constitution, and approval of major financial outlays. Countries with small populations and/or making only small contributions to the running of FAO would be assigned to groups with some commonality, e.g. small Caribbean and small Pacific countries, Central American countries, French-speaking countries of the Sahel. All member countries would be invited to send representatives to the FAO governing conference; the groups of small countries would determine which one spoke and voted for them.

Restriction of the number of entitlements to vote for a director general candidate, and on the FAO budget, would reduce the number of representatives attending and so feeling they ought to speak. The biennial conference would then become a much more efficient and significant occasion.

Ministers of agriculture and other influential government officers are only able to spend a few days in Rome. To make the best use of their time, general statements could be dropped. They will make more impact commenting on specific FAO programmes and performance. Sessions chairpersons can limit the length of individual speeches to ten minutes or less, in the general interest. Governments not sending a delegation would have other opportunities to transmit their views. The overall target would be to reduce the duration of the governing conference and council sessions by at least 50 per cent, so effecting important savings in representation time and servicing costs.

Abolish elaborate representational structures

The FAO regional offices are an anachronism. As a means of contact with the governments of developing countries they have been replaced by the now-extensive coverage of country representatives; these are a direct channel. It is claimed that the regional representatives have a 'political' role. If so, it is not easily seen. They have mobilized support both for existing directors general seeking re-election, and for themselves as candidates. Recent experience in the Near East, where the Cairo office had to be closed for a number of years, has shown that subject matter specialists assigned to regional projects can work from a base in a convenient country representative office. Some regional officers have proved very effective; many have not. Preferable to institutionalizing posts that must then be filled would be an arrangement whereby FAO technical units and others could propose specific regional projects and candidates to implement them, to be judged on their merits and funded on a fixed term basis only.

As early as 1965 an FAO review team recommended that its regional units be withdrawn to headquarters to handle project operations. DG Sen resisted this on the grounds that the regional commissions of the United Nations would then enter FAO's domain. In fact, they set up agriculture divisions to which FAO appointed a further set of staff. This provision can also be abandoned. Instead, FAO can provide information as a matter of course and collaborate in specific joint projects where justified.

Restore the prestige of the technical departments

To emphasize their pre-eminence in an organization concerned with food, agriculture, and rural people, the post of ADG should be reserved for the heads of the agriculture, economic and social, fisheries, and forestry departments. The ADG posts in non-technical units can be suppressed, together with their assistants and secretaries. The strength of FAO lies in its professional competence, so its highest posts should be reserved for leaders in its main fields of work.

Director General Saouma has shown an admirable interest in filling senior posts with staff who have worked in the organization and shown themselves capable. Periodically, however, change in technical leadership can be healthy. There should be a selection from people who are recognized as outstanding, who have specific ideas or experience to contribute. The practice of inviting influential member governments to nominate a candidate has sometimes resulted in the appointment to high posts of ineffective personalities.

ADG posts in development, general affairs, and programme and budget have been created to reward specific incumbents, or to gratify governments in return for a vote. The liaison, protocol, and similar responsibilities associated with general affairs and development ADG posts can be handled at a lower level. Important issues would in any event be referred to the DG's office for a decision. Necessarily, the

Director of Information must be a professional. He should be able to exercise considerable discretion, with resort to the ADGs on matter in their subject areas, and to the DG if his clearance is deemed necessary.

If there is a deputy director general, there is no need for a high-level chief of programme and budget, or a chef de cabinet (who is closer than either to the director general). Only one of them is needed to propose and defend a programme of work to FAO's governing conference and its committees. In recent years, the post of deputy director general has been confined to limited specific assignments, or has been kept vacant. The ability of one person to combine both programme and budget and DDG responsibilities has been amply demonstrated by DDGs Wells and West. The programme and budget post could then return to director level. The DG's office would return to the position pre-Saouma, i.e. with a senior general service personal assistant to re-direct correspondence and enquiries according to pre-established criteria. Senior assistants might be appointed for special duties for limited periods, but not to constitute a link in the channel from ADG to DDG to DG.

Sharpen staffing policies for morale and efficiency

The advantages of a young organization are evident in IFAD. Its specific identity is nurtured at the highest level. Its professionals sense a mission; they feel they are breaking new ground. Its support staff are not predominantly Italian – diversity of national origin leaves room for development concern as a common factor.

Could FAO give notice of termination to its present staff and allow small teams in each work area to select the staff they wished to retain, and fill other vacancies by new recruitment? This procedure was followed by the World Bank in the late 1980s when its staff was judged too large. Under financial pressure FAO also reduced its staff, but by not filling posts when they became vacant, and by not extending short-term contracts. This resulted in sharp imbalances of staff and workload between one unit and another. It also lowers staff quality – those who have nowhere else to go accumulate seniority and are the last to leave. This is in accord with the soft option preference of FAO administration. Passengers are carried indefinitely, to the detriment of work performance, rather than facing the cost of paying them off. Staff supervisors in FAO have often been pressed by personnel to retain or find a place for staff members who were not worth their salary. Their evident inadequacy, or low work contribution, had a negative effect on staff morale around them. When appointed to field advisory assignments they lowered the reputation of FAO's technical assistance. Often, such people were kept on to mandatory retiring age. Early retirement has, however, become fashionable. Many such staff members would welcome this and be ready to compromise on a financial settlement.

Clearly there has to be protection against any one unit becoming staffed by co-nationals or friends of its chief. Where, however, 'geographical' considerations require the appointment of a person whose suitability is not proven, can this be for a fixed term from the outset so that the chief of the unit is not obliged to show outright incompetence to free the post? Transfer to another post more suited to such a person's abilities would still be open. Posts established or upgraded to reward a specific person, or to free his or her previous post for a new incumbent, should disappear, when their immediate purpose is served. Often they have been continued on the specious argument that if they had been established they must be needed.

Full accountability of field projects

As an assurance to agencies and voluntary groups implementing projects through FAO, and to provide more pressure for efficient performance, all FAO field projects should be accountable. This is a necessary check on counter-pressures to get the money spent on time and to maintain smooth relations with the recipient government. Periodic independent inspection can only be salutary. Aid projects should not be shielded from this by the continuing presentation to public opinion of underfed children and images of misery.

Along with the implementation of these measures will come a reduction in FAO staff. The productive technical divisions and technical assistance support teams should stand out in a clearer focus and with more prestige. The criteria of professional competence and effective project implementation will return to preeminence. Internal efficiency will benefit from closer contacts within smaller units. FAO will again attract the best people in its fields of work and merit full support.

REFERENCES

Abbott, J.C. (1958) *Marketing Problems and Improvement Programmes*, Rome: FAO.
────── (1968) 'Marketing issues in agricultural development planning', in F. Moyer and S.G. Hollander (eds) *Markets and Marketing in Developing Countries*, Homewood, Ill.: Irwin.
────── (1986) *Marketing Improvement in the Developing World*, Rome: FAO.
────── (1988) *Agricultural Processing for Development*, Aldershot: Gower.
Beigbeder, Y. (1988) *Threats to the International Civil Service*, London: Pinter Publications.
Bennet, H.L. (1988) *International Organisations: Principles and Issues*, Englewood Cliffs, N.J: Prentice Hall.
Bowbrick, P. (1988) *Practical Economics for the Real Economist*, London: Graham and Trotman.
Boyd Orr, J. (1966) *As I Recall*, London: MacGibbon & Kee.
Cockcroft, L. (1990) *Africa's Way: A Journey from the Past*, London: I.B. Tauris.
Drucker, P. (1958) 'Marketing and economic development', *Journal of Marketing* 22 (1): 252–9.
FAO (1972) *The Role of the Entrepreneur in Agricultural Marketing Development*, Rome: FAO.
────── (1987) *Programme of Work and Budget for 1988–89*, Rome: FAO.
────── (1988) *WCARRD Ten Years of Follow-up: the Impact of Development Strategies on the Rural Poor*, Rome: FAO.
Feld, W.J. and Jordan, R.S. (1988) *International Organisations: a Comparative Appraisal*, New York: Praeger.
Fenn, M.G. (1977) *Marketing Livestock and Meat*, 2nd edition, Rome: FAO.
Gill, P. (1986) *A Year in the Death of Africa*, London: Paladin Grafton Books.
Hall, D.W. (1970) *Handling and Storage of Food Grains in Tropical and Subtropical Areas*, Rome: FAO.
Hambidge, G. (1955) *The Story of FAO*, New York: van Nostrand.
Hancock, G. (1989) *Lords of Poverty*, London: Macmillan.
Harper, M. and Kavura, R. (eds) (1982) *The Private Marketing Entrepreneur and Rural Development*, Rome: FAO.
IFAD (1990) 'Report of brainstorming meeting on the state of world rural

REFERENCES

poverty 26–27 March 1990', Rome: IFAD.

Lele, U. (1989) Review of 'Marketing improvement in the developing world', Rome: FAO, 1986, in *Agricultural Economics* 13 (2): 104–65.

Linnear, M. (1985) *Zapping the Third World*, London: Pluto Press.

Mittendorf, H.J. (1968) 'Marketing aspects in planning agricultural processing enterprises in developing countries', *FAO Monthly Bulletin of Agricultural Economics and Statistics* 17 (4): 1–8.

—— (1976) *Planning of Urban Wholesale Markets for Perishable Food*, Rome: FAO.

Phillips, R.W. (1981) *FAO: its Origin, Formation and Evolution, 1945–1981*, Rome: FAO.

—— (1986) *The World was my Barnyard*, Parsons, West Virginia: MacClain Printing Co.

Schubert, B. (1982) *Agricultural Market Information Services*, Rome: FAO.

Sen, B.R. (1982) *Towards a Newer World*, Dublin: Tycooly.

Shefrin, F. (1980) 'The agriculture agencies; objectives and performance', *International Journal* 35 (2): 38–56, Canadian Institute of International Affairs.

Siamwalla, A. (1975) 'Farmers and middlemen: aspects of agricultural marketing in Thailand', Bangkok: UN Asian Development Institute.

Spinks, G.R. (1970) 'Attitudes towards agricultural marketing in Asia and the Far East', *FAO Monthly Bulletin of Agricultural Economics and Statistics* 19: 1–9 Rome: FAO.

Talbot, R.B. (1982) 'The four world food organizations', *Food Policy*, August, 207–21.

Talbot, R.B. and Moyer, H.W. (1987) 'Who governs the Rome food agencies?', *Food Policy*, November, 349–64.

Von Braun, J., Hotekkiss, D., and Immink, M. (1989) *Nontraditional Export Crops in Guatemala: Effects on Production, Income and Nutrition*, Washington: IFPRI.

Wickstrom, B. (1971) *Marketing of Protein-Rich Foods in Developing Countries* New York: Protein Advisory Group.

Wierer, K. and Abbott, J.C. (1978) *Fertilizer Marketing*, Rome: FAO.

Williams, D. (1987) *The Specialized Agencies and the UN: the System in Crisis*, London: Hurst.

Yeats, P.L. (1955) *So Bold an Aim*, Rome: FAO.

INDEX

de Farcy, H, 180
Faunce, A.D. 56
Felsonvanyi, N. 55
fertilizers 3, 8, 27, 34, 59, 123, 140
 144, 157–8
Finn, D.B. 55
Finney, R.S. 61–2
fish 126, 132, 136, 138
food aid 30
food crops 31, 135, 140, 144, 193
Food for Work 123, 129, 169
food security 185
France 102
fruit 147–51

Galpin, S.L. 173, 176, 186
Geneseth, F. 61
Germany 12, 48, 85, 175, 188
Ghana 32, 132, 135, 140
Gill, P. 28
Gini, E. 58
Glesinger, E. 45–6, 55, 59, 66
governments 95, 146, 150–73,
 180–2, 189–91, 196–9
grain 154–7
Granieri, G.F. 80
Greece 148
Guatemala 144
Guinea 184

Haines, M. 166
Harrison, K. 187
Heseltine, N. 66
Hemsted, C. 78
Henderson, A. 91
Holsten, G. 148
Holt, S.J. 54
Horning, H.M. 49
Hot Springs Conference 2, 17, 102
Hunter, G. 181
Hyden, G. 181

Idi Amin 50, 63–4
Ignatiev, V. 65–6
ILO tribunal 39
incentives 12, 140, 169–70, 175
India 12, 126, 130, 132, 154 162,
 167, 187
Indonesia 138, 188

Ingram, J. 29
International Development
 Association (IDA) 4, 27–8
International Food Policy Research
 Institute (IFPRI) 8, 32
International Fund for Agricultural
 Development (IFAD) 26, 28, 112
International Institute of Agriculture
 (IIA) 2, 55, 102
International Institute of Tropical
 Agriculture (IITA) 31–2
International Maize and Wheat
 Improvement Centre (CIMMYT)
 31
International Monetary Fund (IMF)
 181, 193
International Rice Research Institute
 (IRRI) 31
International Trade Centre (ITC) 7,
 182
investments 28, 36, 46, 147
Iran 161, 163
Ireland 75–6
irrigation 64, 147, 178, 194
Italy 2, 81, 86, 102, 113–17
Islam, N. 54
Israel 23, 176
Ivory Coast 17

Jackson, Sir Robert 72
Jackson, R.I. 44
Japan 13, 41, 48, 86, 193, 195
Jordan 148, 186

Kaldor, N. 9
Kallay, K. 65
Kenya 64, 121, 130, 137, 144–5, 161,
 168, 177, 182
Kesteven brothers 56
King Victor Emmanuel III 2
Kley, W. 168

Latin America 10, 23, 121–2, 142,
 151–2, 165–6
Lebanon 38, 165
Lee, C.Y. 59–60, 175
Leeks, A.J. 80
Lele, U. 156, 171
Libya 163, 176–7

INDEX